New Research on Neurofilament Proteins

NEW RESEARCH ON NEUROFILAMENT PROTEINS

ROLAND K. ARLEN
EDITOR

Nova Science Publishers, Inc.
New York

Copyright © 2007 by Nova Science Publishers, Inc.

All rights reserved. No part of this book may be reproduced, stored in a retrieval system or transmitted in any form or by any means: electronic, electrostatic, magnetic, tape, mechanical photocopying, recording or otherwise without the written permission of the Publisher.

For permission to use material from this book please contact us:
Telephone 631-231-7269; Fax 631-231-8175
Web Site: http://www.novapublishers.com

NOTICE TO THE READER

The Publisher has taken reasonable care in the preparation of this book, but makes no expressed or implied warranty of any kind and assumes no responsibility for any errors or omissions. No liability is assumed for incidental or consequential damages in connection with or arising out of information contained in this book. The Publisher shall not be liable for any special, consequential, or exemplary damages resulting, in whole or in part, from the readers' use of, or reliance upon, this material.

Independent verification should be sought for any data, advice or recommendations contained in this book. In addition, no responsibility is assumed by the publisher for any injury and/or damage to persons or property arising from any methods, products, instructions, ideas or otherwise contained in this publication.

This publication is designed to provide accurate and authoritative information with regard to the subject matter covered herein. It is sold with the clear understanding that the Publisher is not engaged in rendering legal or any other professional services. If legal or any other expert assistance is required, the services of a competent person should be sought. FROM A DECLARATION OF PARTICIPANTS JOINTLY ADOPTED BY A COMMITTEE OF THE AMERICAN BAR ASSOCIATION AND A COMMITTEE OF PUBLISHERS.

Library of Congress Cataloging-in-Publication Data

New research on neurofilament proteins / Ruby K. Arlen (editor).
    p. ; cm.
Includes bibliographical references and index.
ISBN 13 978-1-60021-396-0
ISBN 10 1-60021-396-0
1. Cytoplasmic filaments. 2. Neuropeptides.   I. Arlen, Ruby K.
[DNLM: 1. Neurofilament Proteins--physiology.   WL 104 N532   2006]
QH603.C95N49   2006
611'.0188--dc22
                                                                                     2006024757

Published by Nova Science Publishers, Inc. ✦ *New York*

Contents

Preface		vii
Chapter I	**Neurofilament Proteins in the Cochlea: Changes in Response to Deafening and Neurotrophin Administration** *A. K. Wise, L. N. Pettingill and R. T. Richardson*	1
Chapter II	**Neurofilament Proteins in Brain Diseases** *Olivier Braissant*	25
Chapter III	**Neurofilament Protein Partnership, Export, Transport, Phosphorylation and Neurodegeneration** *Aidong Yuan*	53
Chapter IV	**Neurofilament Changes in Multiple Sclerosis** *Alastair Wilkins*	81
Chapter V	**The Value of Neurofilament-Immunohistochemistry for Identifying Enteric Neuron Types – Special Reference to Intrinsic Primary Afferent (Sensory) Neurons** *Axel Brehmer*	99
Chapter VI	**Insulin Effects on Neurofilament Phosphorylation** *Ruben Schechter and Kenneth E. Miller*	115
Chapter VII	**Neurofilaments in the Mammalian Visual System: As Revealed by SMI 32 Immunohistochemistry** *Zsolt B. Baldauf*	131
Index		163

Preface

Type III intermediate filament proteins assemble into neurofilaments, the major cytoskelatal element in nerve axons and dendrites. They consist of three distinct polypeptides, the neurofilament triplet. Types I, II, and IV intermediate filament proteins form other cytoskeletal elements such as keratins and lamins. It appears that the metabolism of neurofilaments is disturbed in Alzheimer's disease, as indicated by the presence of neurofilament epitopes in the neurofibrillary tangles, as well as by the severe reduction of the expression of the gene for the light neurofilament subunit of the neurofilament triplet in brains of Alzheimer's patients. This book presents new frontal research in such fields as hearing research, brain disease, multiple sclerosis, immunology, diabetes and vision research.

Neurofilaments are important structural and cellular transportation proteins found in both afferent and efferent auditory neurons and their axons. The peripheral processes of auditory neurons innervate sensory hair cells within the cochlea and a loss of these hair cells, by disease or damage, is a major cause of deafness. Hair cells do not naturally regenerate and their loss limits the ability of the cochlea to convert sound waves into electrical signals for auditory perception. The cochlear implant (Bionic Ear) acts to replace the sensory transduction process by electrically stimulating the auditory neurons directly, to effectively bypass the hair cells. However, hair cell loss can also have profound effects on the auditory neurons themselves as hair cells provide them with factors, for example neurotrophins, that are essential for their survival and normal functioning. Once the hair cells have been damaged there is pronounced demyelination and retraction of the peripheral processes followed by the eventual death of the neuron. As the peripheral processes retract from the damaged hair cells the neural scaffold that is provided by neurofilament proteins undergoes active degradation by digestive proteases via mechanisms that are now beginning to be understood. However, this degenerative process can be stopped, and to a certain extent reversed, by providing the auditory neurons with an alternative source of neurotrophins. In addition, resprouting of the peripheral processes can occur following hair cell damage and this can be augmented by the administration of neurotrophins. Neurofilaments play a central role in the resprouting behaviour, and growth cones of resprouting processes have been shown to contain a dynamic population of neurofilament subunits that may facilitate axonal elongation. During resprouting, neurofilaments provide structural integrity and induce maturation of the peripheral processes by regulating phosphorylation-dependent interneurofilament interactions and also interactions between neurofilament proteins and other cytoskeletal components. Results from *in vivo* research will be presented in chapter 1 to describe neurofilament expression in the normal, deafened and neurotrophin treated cochlea detailing the changes observed during the degeneration of auditory neurons and resprouting of their peripheral processes.

Neurofilaments are the main components of intermediate filaments in neurons, and are expressed under three different subunit proteins, NFL, NFM and NFH. Neurofilaments act with microtubules and microfilaments to form and maintain the neuronal structure and cell shape. Phosphorylation is the main post-translational modification of neurofilaments, which influences their polymerization and depolymerization, and is responsible for their correct assembly, transport, organization and function in the neuronal process. In particular, phosphorylation is essential for the repulsion of the neurofilament polymers in axons, which determines the axonal diameter and the velocity of electrical conduction. The phosphorylation state of neurofilaments is regulated in a complex manner, including interactions with the neighbouring glial cells.

Abnormal expression, accumulation or post-translational modifications of neurofilament proteins are found in an increasing number of described neurological diseases, such as amyotrophic lateral sclerosis, Parkinson's, Alzheimer's and Charcot-Marie-Tooth diseases, or giant axonal neuropathy. Some of these diseases are associated with mutations discovered in the neurofilament genes. Recently, altered expression and phosphorylation states of neurofilament proteins have also been shown in metabolic diseases affecting the central nervous system either during development or in adulthood, such as hepatic encephalopathy due to hyperammonemia, methylmalonic and propionic acidemias, and diabetic neuropathy. Finally, accumulation of neurofilament proteins in the cerebrospinal fluid has been described as discriminating marker for patients with multiple sclerosis, and as predictor of long-term outcome after cardiac arrest. This review in chapter 2 will focus on the most recent investigations on neurofilament proteins in neurodegenerative, neurodevelopmental and metabolic diseases, as well as on the use of neurofilaments as markers of diseases.

Neurofilaments are about 10 nm diameter intermediate filaments of neurons which add rigidity and tensile strength to neurites (axons and dendrites) and determine neurite caliber. They are generally believed to be obligate heteropolymers composed of the neurofilament triplet proteins, designated NF-L, NF-M and NF-H for light, medium and heavy molecular weight subunits, respectively. Two other related intermediate proteins alpha-internexin and peripherin may be closely associated with neurofilament. Recent analyses of mice in which single or multiple neurofilament genes are deleted have defined the minimum structural requirements for efficient neurofilament export and transport and also showed the dissociation of neurofilament number in axons with the transport rate of neurofilament. After their synthesis, neurofilament proteins form proper partnership with one another and are exported into and transported throughout axons primarily in the form of heterooligomers or shot filaments where they are incorporated into or exchanged with a basically stationary neurofilament network. Neurofilament proteins distribute non-uniformly along axons with phosphorylation increasing proximally to distally. The gene deletion analyses and additional mutagenesis studies provide new evidences for us to reconsider the basic composition of neurofilament and how phosphorylation controls neurofilament transport. Surprisingly, deletion of the NF-H or its phosphorylated tail domain or NF-M phosphorylated tail domain does not alter neurofilament transport rate in vivo and, the NF-M tail domain rather than NF-H or its tail domain proved to be essential for determining the radial growth and conduction velocity of large myelinated axons. The observations that mutations of NF-L are a cause of Charcot-Marie-Tooth disease type 2E/1F, a neuronopathy associated with neurofilament accumulation and discovery of neurofilament inclusion disease, a form of frontal temporal dementia with hallmark lesions containing neurofilaments and alpha-internexin, support growing evidence that neurofilament may have a causative role in neurodegeneration. In chapter 3, a model was proposed to explain the role of neurofilament in neurodegeneration, i.e., gene mutation or environmental damage on axons could cause loss of neurofilament protein proper partnership, inefficient export of neurofilament and accumulation of remaining partner in cell boy and proximal axons, resulting in age-related axonal atrophy or induce abnormal movement or retraction of axonal phosphorylated neurofilaments back into cell bodies or axonal spheroids, leading to progressive neurodegeneration and eventually cell death.

Multiple sclerosis is a disease of the central nervous system characterised by episodes of neurological dysfunction which often recover, usually followed some years later by progressive and irreversible decline. Lesions of multiple sclerosis are characterised by varying degrees of inflammation, myelin and oligodendrocyte loss, astrogliosis and axonal pathology. Axonal loss is seen in progressive phases of the disease and appears to correlate well with clinical disability. Prior to loss of axons, pathological specimens have revealed changes in the immunohistochemical phenotype of axons. Specifically there may be evidence of dephosphorylation of neurofilaments within axons and transection of axons leading to the formation of axonal spheroids which are rich in dephosphorylated neurofilaments. Evidence of axonal transport defects may also be found in lesions with accumulation of amyloid precursor protein within the axon. Mechanisms of axonal pathology in multiple sclerosis remain unknown, but it is likely that inflammation causes some degree of damage in the acute phases. However there is evidence that axonal loss may continue even in the absence of inflammation. It has been postulated that axonal loss in this situation occurs due to loss of trophic support from myelin and oligodendrocytes. Oligodendrocytes and myelin are known to provide trophic support for axons and specifically can influence phosphorylated neurofilament levels. The precise mechanisms are unknown, but recent evidence suggests a combination of contact mediated and soluble factors may increase neurofilament phosphorylation and promote axonal protection. Knowledge of such mechanisms may lead to improved therapies to prevent progressive disease. This review in chapter 4 will discuss axonal changes in multiple sclerosis, specifically alteration in neurofilament phosphorylation states, and potential mechanisms of axonal protection.

The enteric nervous system (ENS), arranged in different nerve networks within the wall of the gastrointestinal tract, harbours numerous different neuron types which can be grouped into three main populations according to their functions. Intrinsic primary afferent neurons (IPANs) are the first links of enteric neuronal reflex arches, interneurons form ascending or descending chains of like neurons within the intestinal wall and motor neurons provide output to, e.g., the gut muscle, mucosa or blood vessels. Most of our knowledge of the neuronal components of enteric circuits, including functional evidence of IPANs, is derived from studies in the guinea-pig. An efficient method besides other techniques, to distinguish between functionally different neuron types in a given species is the evaluation of the chemical coding of enteric neurons, i.e. the immunohistochemical proof of the presence and co-existence of neuronal markers mainly within their cell bodies. However, the chemical codes of enteric neurons are species specific. During the last decade, various attempts have been undertaken to identify the functionally equivalent neurons in the human gastrointestinal tract. Combined tracing and immunohistochemical studies revealed projections and chemistries of putative inter- and motor neurons in the human.

In chapter 5, the authors describe the value of immunohistochemical staining for neurofilaments (NF), an important component of the cytoskeleton of subgroups of neurons, for both the identification of and the discrimination between different neuron types, which are visualized by their processal architectures, in particular the putative myenteric IPANs in the humans and pigs. These are morphological type II neurons, i.e., they are non-dendritic and multi-axonal neurons. The multiple axons of type II neurons run both circumferentially, within the plane of the myenteric plexus around the gut lumen, and vertically, penetrating the circular muscle and running to the submucosa and mucosa. They demonstrate that both calbindin, a marker for IPANs in the guinea-pig, and calcitonin gene-related peptide, a marker for IPANs in the pig and mouse, are absent from the majority of myenteric type II neurons in human. Thus, the main, and at this time the only applicable, way of identifying IPANs in the human ENS is their type II-morphology visualized through NF-immunohistochemistry.

In chapter 6, the authors demonstrated the production of neuronal insulin [I(n)] within the fetal brain in vivo and neuron cell cultures. I(n) is present within neuron cell cultures at a concentration of 320 pg/ ml. They demonstrated that I(n) affects neurofilament distribution in neuron cell cultures and promotes axonal growth via mitogen activated protein kinase (MAPK or ERK).

Neurofilaments need to be phosphorylated to be distributed to the axons and form sidearms. Sidearms contribute to neurofilament anchoring, axonal caliber and growth. In diabetes mellitus, neurofilament content has been shown to be decreased. The authors investigated if I(n) promoted neurofilament phosphorylation and the cascade involved. Neuron cell cultures from 19 days gestational age rat brains were incubated in insulin free/serum free medium and neurofilament distribution was studied by immunohistochemistry employing a mouse monoclonal anti-phosphorylated neurofilament antibody. Neurons were treated with different inhibitors for 1 hour: 1) 10 μ M of isoproterenol (an insulin receptor tyrosine kinase inhibitor); 2) 50 μ M of PD98059 (a ERK inhibitor); 3) 25 μ l/ml of a guinea pig anti-porcine insulin antibody, 4) 100 μ M of wortmannin a phosphatidylinositol 3 kinase (PI-3K) inhibitor. After isoproterenol or PD98059 treatments, the neurons were exposed to 5 ng/ml of insulin for 1 hour. Neurons in IFM showed phosphorylated neurofilament along the axon without interruptions. In neurons treated with isoproterenol, anti-insulin antibody or PD98059, the distribution of the phosphorylation changed to a punctuate sporadic localization along the axon. The addition of exogenous insulin to the neuron cell cultures reversed the effects of isoproterenol or PD98059, but not the anti-insulin antibody. Wortmannin did not affect neurofilament phosphorylation. Thus, I(n) has a role in neurofilament distribution and phosphorylation and, as we previously demonstrated, promoting axonal growth.

In the brain of the insulin knockout mouse (I-/-), they demonstrated that the medium molecular weight was hyperphosphorylated ($p<0.05$) when compared to wild type mice. The abnormal neurofilament hyperphosphorylation induced an aberrant assembly and transport of the neurofilament. We also quantitated kinases such as glycogen synthase kinase 3 ß (GSK-3 ß) GSK-3, ERK and c-Jun N-terminal kinase (JNK), that phosphorylate the different motifs of the neurofilament tail domain. In I-/-, we found that JNK 1 and 3 were hyperphosphorylated ($p<0.05$), ERK 1 was decreased in phosphorylation ($p<0.05$) and no changes occurred in GSK-3 when compared to wild type animals. These studies confirm that insulin via ERK phosphorylates neurofilament as shown in neuron cell cultures. Insulin deficiency may decrease the formation of medium molecular weight neurofilament sidearm, axon caliber, axonal transport velocity and, in consequence, cognition as seen today in patients of type 1 diabetes mellitus.

The histochemical detection of neurofilaments (Nfs) has proved to be a powerful tool in demarcating several cortical and subcortical areas in the mammalian visual system. Apart from the Nfs' cortical parcellating power, it delineates previously unequivocal histological boundaries within the associative thalamic and pretectal nuclear complexes. In addition, the accumulation of Nfs is a true marker of large brainstem oculomotor neurons and cortical layer V giant pyramidal neurons alike in various regions. The parallelism of the visual system can easily be detected and observed in the characteristic distribution of Nfs. These polypeptides are strongly expressed alongside a visual stream, the magnocellular (M) pathway, at both cortical and subcortical levels. The faster-conducting M pathway's axons are of a larger caliber, and thus display more cytoskeletal Nfs.

Research on Nfs has gained medical importance lately due to their presumptive role in various neurodegenerative diseases. A monoclonal antibody (SMI 32) readily recognizes the presence of the Nf triplet in mammalian, and also in marsupial, nervous systems. In chapter 7, the immunohistochemical distribution of SMI 32-reactive Nfs along the rodent, carnivore and primate visual system is discussed.

In: New Research on Neurofilament Proteins
Editor: Roland K. Arlen, pp. 1-24

ISBN: 1-60021-396-0
© 2007 Nova Science Publishers, Inc.

Chapter I

Neurofilament Proteins in the Cochlea: Changes in Response to Deafening and Neurotrophin Administration

A. K. Wise[1,2]*, L. N. Pettingill, [1,2] and R. T. Richardson[1,2]

[1]The Bionic Ear Institute, East Melbourne, Australia
[2]Department of Otolaryngology, University of Melbourne, Parkville, Australia

Abstract

Neurofilaments are important structural and cellular transportation proteins found in both afferent and efferent auditory neurons and their axons. The peripheral processes of auditory neurons innervate sensory hair cells within the cochlea and a loss of these hair cells, by disease or damage, is a major cause of deafness. Hair cells do not naturally regenerate and their loss limits the ability of the cochlea to convert sound waves into electrical signals for auditory perception. The cochlear implant (Bionic Ear) acts to replace the sensory transduction process by electrically stimulating the auditory neurons directly, to effectively bypass the hair cells. However, hair cell loss can also have profound effects on the auditory neurons themselves as hair cells provide them with factors, for example neurotrophins, that are essential for their survival and normal functioning. Once the hair cells have been damaged there is pronounced demyelination and retraction of the peripheral processes followed by the eventual death of the neuron. As the peripheral processes retract from the damaged hair cells the neural scaffold that is provided by neurofilament proteins undergoes active degradation by digestive proteases via mechanisms that are now beginning to be understood. However, this degenerative process can be stopped, and to a certain extent reversed, by providing the auditory neurons with an alternative source of neurotrophins. In addition, resprouting of the

---

* Corresponding author: Dr. Andrew Wise, awise@bionicear.org

peripheral processes can occur following hair cell damage and this can be augmented by the administration of neurotrophins. Neurofilaments play a central role in the resprouting behaviour, and growth cones of resprouting processes have been shown to contain a dynamic population of neurofilament subunits that may facilitate axonal elongation. During resprouting, neurofilaments provide structural integrity and induce maturation of the peripheral processes by regulating phosphorylation-dependent inter-neurofilament interactions and also interactions between neurofilament proteins and other cytoskeletal components. Results from *in vivo* research will be presented to describe neurofilament expression in the normal, deafened and neurotrophin treated cochlea detailing the changes observed during the degeneration of auditory neurons and resprouting of their peripheral processes.

Introduction

The function of the cochlea is to transduce sound waves into electrical stimuli for transmission of auditory information to higher brain centres. Central to this function are the auditory neurons that innervate the sensory regions of the cochlea and provide both efferent and afferent nerve fibres. Sensorineural hearing loss is a common pathology of the auditory system in which the sensory hair cells are damaged by infection, exposure to loud noise, physical trauma, pharmacological agents, ageing or genetics. Damaged hair cells do not naturally regenerate and once lost cause a secondary neural degenerative process that can ultimately lead to the death of the auditory neurons. Current, and also future, therapies that may be used to restore hearing to deaf people, such as the cochlear prosthesis, require a viable population of auditory neurons. Therefore it is of paramount importance to prevent and even reverse the degenerative process that follows deafening to ensure that the remaining auditory neurons are capable of transmitting auditory information necessary for hearing.

Neurofilaments are a family of proteins that are found in auditory neurons and are integral to their normal functioning. Neurofilaments are a type of intermediate filament expressed almost exclusively in neuronal cells. They are the dominant proteins of the neuronal cytoskeleton and act together to form and maintain cell shape, provide rigidity and tensile strength, and facilitate the transport of particles and organelles within the cytoplasm. Neurofilaments are also thought to be directly involved in the control of the radial growth of axons, defining axonal calibre and contributing to cytoskeletal stability.

The prominence of neurofilaments in auditory neurons makes them a useful marker of neuron damage and regeneration. Experiments discussed here have exploited the expression profile of neurofilaments to examine pathological changes of auditory neurons following sensorineural hearing loss and changes following the administration of neurotrophins into the cochlea. We have observed alterations in neurofilament expression that accompanies the degeneration and demyelination of the auditory neurons. Results have also confirmed that neurofilaments are key structural proteins found in the resprouting peripheral processes of auditory neurons.

The Cochlea

The cochlea is a spiral-shaped, bone-encapsulated organ that contains two cell types important for the transduction of sound to the brain; *sensory hair cells* and *auditory neurons* (Figure 1A). The hair cells reside in a structure known as the organ of Corti that lies above the basilar membrane, while the auditory neurons reside in Rosenthal's canal and send peripheral processes that terminate or synapse with hair cells. Sound waves cause the basilar membrane to vibrate in frequency-specific locations along the spiral, with hair cells near the basal end of the cochlear spiral responding to high frequency sound and hair cells near the apical end of the spiral responding to low frequency sound. Vibration of the basilar membrane initiates the release of neurotransmitter from hair cells, which bind to receptors on the peripheral processes causing depolarization and the subsequent generation of an action potential. This signal is then transmitted via a number of brainstem pathways to the auditory cortex.

The organ of Corti contains two types of sensory hair cells (Figure 1A). The *inner hair cells* receive afferent innervation from large myelinated Type I neurons that comprise 90-95% of all sensory afferents (Spoendlin, 1972). The Type I neurons are commonly known as spiral ganglion neurons (SGNs), and are bipolar neurons with relatively thick 1-2μm diameter processes. Smaller unmyelinated Type II afferent neurons innervate the *outer hair cells* and comprise the remaining 5-10% of the SGN population. Efferent neurons are either large myelinated medial neurons that innervate the outer hair cells, or small unmyelinated lateral neurons that form presynaptic terminals with the peripheral processes of the SGNs at the base of the inner hair cells (Spoendlin, 1972; Guinan et al., 1983). Both afferent and efferent neurons extend their peripheral processes through the osseous spiral lamina (OSL), with the myelinated neurons losing their myelin sheath at the habenula perforata before entering the organ of Corti as unmyelinated endings and forming synaptic connections with their target (Figure 1A).

Neurofilaments are major cytoskeletal components of auditory neurons and are present during development and in the adult cochlea (Dau and Wenthold, 1989; Hafidi et al., 1990; Romand et al., 1990; Berglund and Ryugo, 1991). In the adult guinea pig cochlea neurofilaments are prominent in the peripheral processes (Figure 1B) as they project through the OSL towards the organ of Corti. They are also prominent within the cell bodies of the SGN themselves where they link together in a dense network to provide the cell bodies with structural support (Figure 1C).

Neurofilament Structure, Assembly and Transport

Neurofilaments are members of the highly conserved intermediate filament family of cytoskeletal proteins. Based on molecular weight, neurofilaments are classified as low (NF-L; 68kD), medium (NF-M; 125kD), and high (NF-H; 200kD) (Pachter and Liem, 1984; Scott et al., 1985; Schlaepfer, 1987). In most mature mammalian neurons the three neurofilament subunits assemble as triplet proteins forming a 10nm filamentous structure that runs the

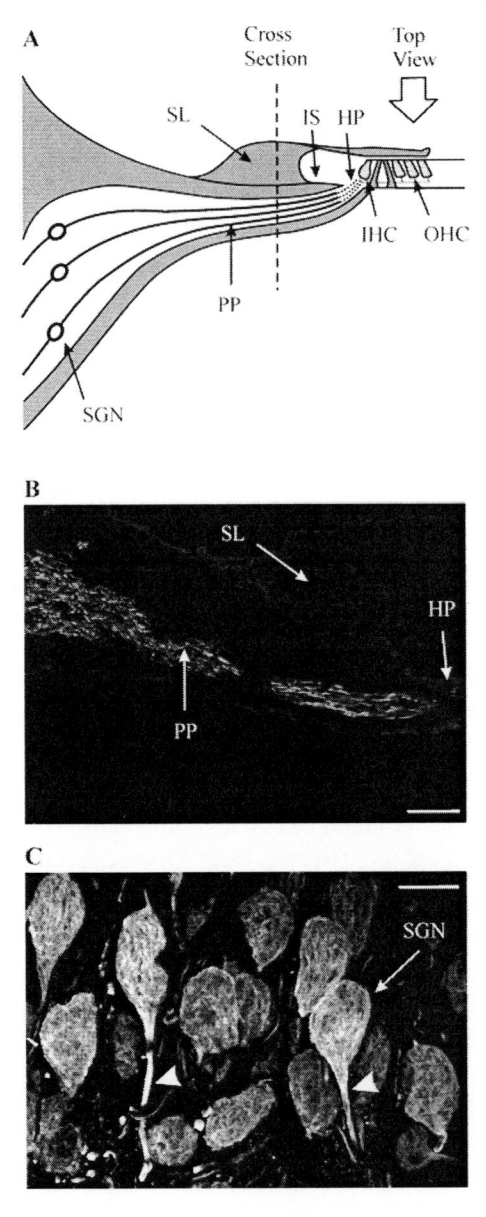

Figure 1. A Schematic diagram of a cochlear section showing the spiral ganglion neurons (SGNs), the spiral limbus (SL), the inner sulcus (IS), the habenula perforata (HP), and the inner and outer hair cells (IHC, OHC). The cell bodies of the SGNs (arrow) are located in Rosenthal's canal and extend their myelinated peripheral processes (PP) through the osseous spiral lamina (OSL). The peripheral processes become unmyelinated at the habenula perforata (dotted lines) before synapsing with the inner and outer hair cells. Cross-sections through the OSL (dashed line) were made in order to count the peripheral processes and measure their diameter (see figure 4). Whole mounts were constructed to enable a 'top view' of the OSL (see figure 2). B A thin section (5μm) of a normal guinea pig cochlea stained for high molecular weight neurofilament (NF-H). The NF-H protein is expressed in the myelinated peripheral processes (scale bar 50μm). C NF-H staining of SGNs in a thick section (50μm) from a normal guinea pig cochlea. The NF-H protein was expressed in the cell bodies of the SGNs (arrow) in a lattice-like network of filaments that provide structural support to the cell. Two myelinated central processes are prominent (arrow heads) (scale bar 20μm). (A adapted from Wise et al., 2005).

length of neuronal axons (Hoffman and Lasek, 1975). NF-M and NF-H also have long, flexible polypeptide chains, called sidearms, which emanate from the surface of the axial core of the C-terminal tail domains (Angeletti et al., 1985). The flexible sidearm chains link the neurofilaments to each other and to other elements in the cytoskeleton, for the formation of extremely stable networks and bundles. Assembly of neurofilament chains is a multi-step process involving interactions between phosphorylated NF-M and NF-H and other cytoskeletal proteins. Neurofilaments are produced in the cell body and transported down the axon as either polymeric structures (the structural hypothesis) or as individual subunits or small oligomers (subunit hypothesis) (reviewed by Lariviere and Julien, 2004). Transportation occurs bi-directionally at conventional rates within the axon (up to about 1μm per second), but movement is interrupted by prolonged pauses (>99% of time pausing), so that overall movement is relatively slow (Roy et al., 2000; Wang et al., 2000).

Developmentally, the neuronal cytoskeleton in the SGNs, and indeed in all neuron types, must adapt from the plastic form of early migrating axons, to the stable supporting lattice of functioning mature neurons. In order to accommodate this complex transition, expression of the three neurofilament subunits occurs at specific developmental stages. Firstly, NF-L expression is triggered by neuronal differentiation; shortly thereafter, NF-M expression occurs in conjunction with neurite formation; finally, in the latter stages of maturation, expression of NF-H is associated with radial growth of the axons (Liu et al., 2004).

Neurofilaments and Cochlear Development

Cochlear development begins at embryonic day (E) 7.5 in the mouse as an ectodermal thickening which forms a structure called the otic placode. As cells within the otic placode proliferate, they begin to invaginate to form the otic cup (E8-9) and then pinch off to form an enclosed otic vesicle (E9.5). Precursors of the cochlear (and vestibular) ganglia delaminate from the otic vesicles at the same time as complex morphological changes occur, to result in the fluid-filled cochlear labyrinth by E18. However, the basic architecture of the inner ear is established by E14.5. Hair cells first arise in the developing mouse cochlea around E12-18 with those in the apical turn appearing first (E12) and those in the basal turn appearing last (E16). In contrast, auditory neurons are present in the basal turn at day E12 while those in the apex appear later at day E14-16 (Ruben, 1967).

In the developing mouse cochlea, nerve fibres and soma expressing NF-M are present in the basal and apical turns at E19 (Hasko et al., 1990). In the developing rat cochlea at embryonic days 16, 18 and 20, the peripheral and central processes of SGNs express all three subunits of neurofilaments, while in the soma, expression is restricted to NF-L and NF-M. At 20 days gestation, a distinct population of nerve cells emerges in the basal turn with strong expression of all three neurofilament subunits. This population, representing approximately 6-7% of total SGNs, appears in all turns by 2-3 days after birth. The strong immunoreactivity of this selective neuronal population suggests that they correspond to the Type II SGNs. These results imply that the differences in the neurofilament profile of the Type I and the Type II SGNs occurs perinatally in the rat (Hafidi and Romand, 1989; Romand et al., 1990) although differences in neurofilament expression in the mature cochlea may exist between species (Dau and Wenthold, 1989). A similar phenomenon was encountered in a 14-week gestation human cochlea in which an estimated 20% of the SGN cell bodies reacted intensely with neurofilament antibodies while the remaining neurons reacted more weakly (Anniko et al., 1987). In the same study, it was also noted that there was intense neurofilament staining in nerve fibres even when the corresponding cell body was unlabelled, demonstrating that neurofilaments are not equally distributed within the neuron at this stage of human development.

In the cochlea, neurofilament expression is not restricted to the neurons. NF-M has been detected in sensory hair cells during development with diminishing expression during post-natal stages. In the mouse, NF-M positive hair cells were detected in the basal turn at E19. At post-natal day 1 this expression extended to hair cells of the middle turn, and to the apical turn by day 2. At day 10, outer hair cells had reduced NF-M expression and at post-natal day 21, neither inner or the outer hair cells expressed NF-M (Hasko et al., 1990). A similar temporal gradient of NF-M expression in hair cells was found in the chick cochlea, in which expression began in the mid-frequency range (basal) and extended to the low frequency range of hair cells (apical) later during development (Oesterle et al., 1997). The reduced NF-M expression in hair cells coincides with the onset of auditory function. In a human cochlea after a 14-week gestation period, there was no neurofilament immunoreactivity in hair cells, despite neurofilament immunoreactivity in adjacent nerve fibres (Anniko et al., 1987). The reason for the transient neurofilament expression in hair cells is unknown, but it refutes the theory that neurofilament expression occurs only in neuronal cells. It may be speculated that hair cells require a stage of structural stability during development but then require more flexibility when auditory function begins.

The expression of neurofilaments in the cochlea mirrors the reported development of neurofilaments in other areas of the nervous system. In the rat, NF-L and NF-M expression was detected at approximately E12 as neurites differentiate, and even phosphorylated forms of NF-M could be detected at this early stage. NF-H was detected at E15, also in both phosphorylated and unphosphorylated forms (Carden et al., 1987). The ratio of neurofilament subunits during embryogenesis appears to be important for development of the nervous system. Low levels of NF-H and high levels of NF-L and NF-M early in development promotes elongation and radial growth of neuronal processes, while higher levels of NF-H in later stages of development promotes stability of the nervous system (Kong et al., 1998).

In addition to their role in cochlear development, neurofilaments are also important cytoskeletal proteins in the adult mammalian cochlea. The cell bodies of Type I SGNs and their peripheral processes are immunoreactive to the NF-H subunit only, whereas the Type II afferents express all three of the neurofilament subunits (NF-L, NF-M and NF-H) (Dau and Wenthold, 1989; Berglund and Ryugo, 1991). We have used antibodies to the 200kDa neurofilament subunit (NF-H) to selectively label both the afferent and efferent auditory neurons and observe their innervation pattern in the cochlea.

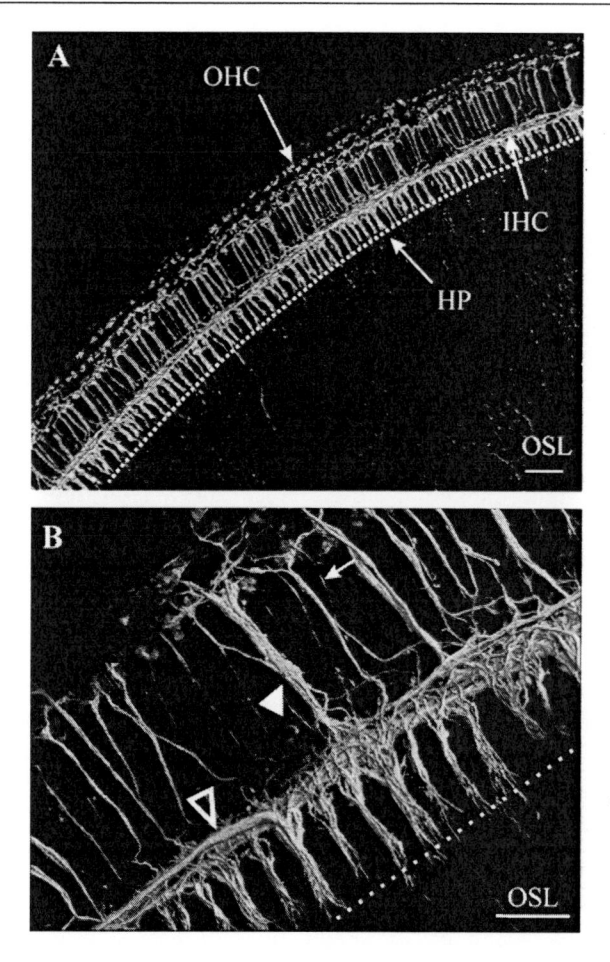

Figure 2. Whole mount preparations showing a 'top view' (see figure 1) of the peripheral processes in the basal region of a normal guinea pig cochlea. Image A shows the NF-H stained unmyelinated region of the peripheral processes as they exit the OSL and enter the organ of Corti (not stained). The myelinated regions of the peripheral processes located within the OSL were largely unstained. The unmyelinated regions of the peripheral processes project through the habenula perforata (HP) and make synaptic connections with the inner and outer hair cells (IHC, OHC; not visible but positions indicated by arrows) (scale bar 50μm). Image B is a higher magnification of the same cochlear region and shows NF-H stained peripheral processes at greater resolution. Peripheral processes were observed to exit individual habenula perforata (dotted line) and project towards the base of the inner hair cells (not visible). Efferent peripheral processes were seen to project laterally along the inner spiral bundle (open arrow head) or cross the tunnel of Corti (closed arrow head). Smaller Type II afferents were also observed to cross the tunnel of Corti (arrow) (scale bar 20μm) (Figure adapted from Wise et al., 2005).

Neurofilaments in the Adult Cochlea

In order to visualize the peripheral processes of the auditory neurons cochlear whole mounts were prepared and processed for confocal fluorescent microscopy enabling 3-dimensional reconstructions of the cochlear tissue (Figure 2, for details see Wise et al., 2005). In the normal adult cochlea, Type I afferents were seen to exit individual openings at the habenula perforata in bundles of 20-30, where they projected towards the organ of Corti to innervate the inner hair cells located nearest to their habenula opening (Figure 2B – hair cells not visible). Efferent processes were observed to project laterally along the inner spiral bundle or to cross the tunnel of Corti towards the outer hair cells (Figure 2B). The tunnel-crossing efferents and Type II afferents were prominently labelled with the NF-H antibody and were distinguished from each other by differences in their size and by the relative position within the organ of Corti (Figure 2B). Type II afferents were smaller and crossed the floor of the basilar membrane embedded in the foot processes of the pillar cells, whereas the efferents travelled midway between the basilar membrane and the reticular lamina.

In cochlear whole mounts, neurofilament staining was prominent in the peripheral processes of the SGNs but was restricted to the most peripheral regions, beginning at the foramen nervosum in the habenula perforata, where the peripheral processes became unmyelinated before projecting towards the organ of Corti. The myelinated regions of the peripheral processes within the OSL were essentially unstained (Figures 2A, B below dotted line). The neurofilament expression in the whole mount preparations was in stark contrast to its expression profile in cochlear cross sections (Figure 1B) in which clear neurofilament expression was observed along the entire length of peripheral processes within the OSL. The different neurofilament expression profile in the two preparations is due to the fact that binding of the neurofilament antibodies to the intact peripheral processes in the whole mount preparation was restricted by the myelin sheath surrounding the processes (Romand et al., 1990; Fechner et al., 1998). This is highlighted by the observation that one small Type II unmyelinated neuron was stained with the NF-H antibody within the OSL (see Figure 2B - below dotted line). In contrast, the cutting of thin (5μm) cochlear sections shown in Figure 1B exposed the previously encased neurofilament cytoskeletal proteins to the antibody. We have taken advantage of this characteristic neurofilament staining profile to document changes to the myelin sheath of the peripheral processes and the peripheral processes themselves in response to sensorineural hearing loss.

Neurofilaments and Neuronal Degeneration

In Wallerian degeneration (Waller, 1850), the axoplasm in the distal stump of the transected or crushed axon undergoes a process of granular disintegration of the cytoskeleton, where a number of axonal proteins, including neurofilaments, are subjected to proteolysis and are converted into an amorphous and granular material (Hall and Lee, 1995). Following nerve crush or section there is an increase in intracellular calcium (LoPachin and Lehning, 1997; Martinez and Ribeiro, 1998) which triggers cytoskeletal breakdown via activation of calpains (calcium-activated neural proteases) present in the nervous tissue (Schlaepfer, 1987; Marques et al., 2003). The calpains rapidly degrade neurofilaments, other cytoskeletal proteins and the myelin sheath, and thereby play a fundamental role in neuron degeneration (Kamakura et al., 1983). Since dephosphorylation of neurofilaments enhances their susceptibility to degradation by calpain (Pant, 1988), it has been suggested that phosphorylation may regulate the stability of neurofilaments (Terada et al., 1998). In addition, it has been suggested that the recycling of the neurofilament degradation products may provide a feedback mechanism to regulate neurofilament production (see Liu et al., 2004). Axon degeneration may involve an active and regulated self-destruction program, rather than a passive 'wasting away'. It is likely that the degenerative processes described here may also occur in cochlear neurons following sensorineural hearing loss. Results from *in vitro* experiments have supported the notion that a common mechanism may underlie different models of axon degeneration (Finn et al., 2000).

Neurofilaments in the Deafened Cochlea

The loss of auditory hair cells by over stimulation (noise trauma) or by pharmacological agents (antibiotics such as neomycin or kanamycin) can cause degeneration of auditory neurons that is secondary to the initial cochlear insult and is called sensorineural hearing loss (Ylikoski et al., 1974; Puel et al., 1995). It is thought that loss of neurotrophic support, provided to the auditory neurons by the hair cells, in particular the inner hair cells, is a major factor in this process (Lefebvre et al., 1992).

We have used an ototoxic deafening model to initiate sensorineural hearing loss and the subsequent degeneration of auditory neurons (Webster and Webster, 1981; Hardie and Shepherd, 1999; Shepherd and Hardie, 2001). The aminoglycoside kanamycin, in combination with the loop diuretic frusemide, was used to produce degeneration of the organ of Corti followed by retrograde degeneration of the auditory neurons in pigmented guinea pigs (n=33) (Wise et al., 2005). Two time periods of deafening were used; a short-term period (33 days) and a long-term period (61 days). In both cases, deafening was effective in eliminating hair cells in all but the most apical regions of the cochlea where some remaining hair cells were observed. There was a significant reduction in the density of SGNs in response to deafening, with the degenerative effects progressive over time, a finding consistent with previous studies (Webster and Webster, 1981; Leake and Hradek, 1988).

Although we did not carry out a systematic examination of changes in NF-H expression in the SGN cell bodies, NF-H was observed in the deafened cochlea (Figure 3D). A previous study that examined the changes in neurofilament expression following deafening with neomycin found that the expression of NF-L and NF-M increased in the SGN cell bodies two to three weeks after sensorineural hearing loss, whereas there was little change in the immunoreactivity of the NF-H subunit over this time period (Dau and Wenthold, 1989). Subsequently, 14-15 weeks after deafening there was a marked decrease in the level of staining for all neurofilament subunits in the remaining SGN cell bodies (Dau and Wenthold, 1989). Changes in the neurofilament expression were more prevalent in the peripheral processes of the auditory neurons than in the cell bodies.

Two methods were employed in order to examine the effects of sensorineural hearing loss on the peripheral processes. As before, cochlear whole mounts were generated and stained with an antibody to NF-H. We have used the expression profile of NF-H as an indicator for degeneration of the peripheral processes and as a useful marker for demyelination of these processes. The second method used cochlear cross sections (see Figure 1A) that were either stained for NF-H or by osmium tetroxide to quantify changes in the myelinated regions of SGN peripheral processes.

Cochlear whole mounts of NF-H stained peripheral processes in the lower basal region of a cochlea deafened for a period of 33 days are shown in Figure 3 A, B and C. There was a dramatic disruption to the normal innervation pattern of the organ of Corti. In particular, there was a significant loss of peripheral processes that cross the basilar membrane and project towards the outer hair cells. There was also a noticeable reduction in the number of peripheral processes that project through the habenula perforata, indicating the loss of Type I afferents. The peripheral processes of the Type I afferents are usually the most numerous neuron type within the habenula perforata and a reduction in the number of these processes was a significant feature of sensorineural hearing loss. In addition, neurofilament labelling of the remaining peripheral processes was observed within the OSL, indicating demyelination of these processes, as labelling of this region was rarely observed in whole mounts of the normal cochlea (Figure 2A,B). Consistent with the disintegration of the neuronal cytoskeleton is the observation of fragments of neurofilaments most likely representing neurofilament breakdown products within the OSL (Figure 3B).

Large accumulations of neurofilament proteins were observed at the distal ends of some peripheral processes in the deafened cochlea (see Figure 3B, C). It was unlikely that these dense neurofilament bodies were growth cones of resprouting processes (shown in Figure 4B) based on their relatively large size and spheroid shape. We can only speculate as to their role in the degeneration process but accumulations of neurofilaments have been reported in neuron degeneration and pathology in some disease states. Neurofilament accumulations are found in the cell bodies of specific neurons in several human neurological diseases, such as neurofibrillary tangles in Alzheimer's disease (Cork et al., 1986) and the Lewy bodies (Forno et al., 1986) and spheroids found in amyotrophic lateral sclerosis (ALS) (Delisle and Carpenter, 1984). Mutations of neurofilament genes have long been sought in ALS because hyper-phosphorylated neurofilament accumulations are such a prominent feature of this disease. Over-expression of neurofilament proteins in transgenic mice can provide a model of ALS or other neurodegenerative diseases, but the precise mechanisms by which neurofilaments accumulate in neurodegenerative disease are nevertheless unclear. The major culprits appear to be abnormalities in neurofilament synthesis, transport, phosphorylation and proteolysis (Liu et al., 2004). For instance, neurofilaments are synthesised in cell bodies and then transported into and through axons, and a disruption of this process has been described in several transgenic models of ALS. Thus, a disruption of axonal transport might underlie the neurofilament accumulation in the cell bodies or the proximal axons in these particular models. However, in the cochlea, neurofilament accumulations were observed at the distal end of the peripheral processes and were not evident in the cell bodies of SGNs. It has been reported that neurofilament synthesis can also occur in axons, perhaps to accommodate rapidly to local needs; such as an immediate response to reassemble and sustain the axonal cytoskeleton during nerve regeneration (Grant and Pant, 2000). Since auditory neurons have been reported to spontaneously regenerate following sensorineural hearing loss (for example, see Lawner et al., 1997; Wise et al., 2005), it is possible that the neurofilament accumulations observed in the peripheral processes represent plasticity of these injured neurons and may indicate abortive attempts of regeneration of these peripheral processes (for example see Shintaku et al., 1996).

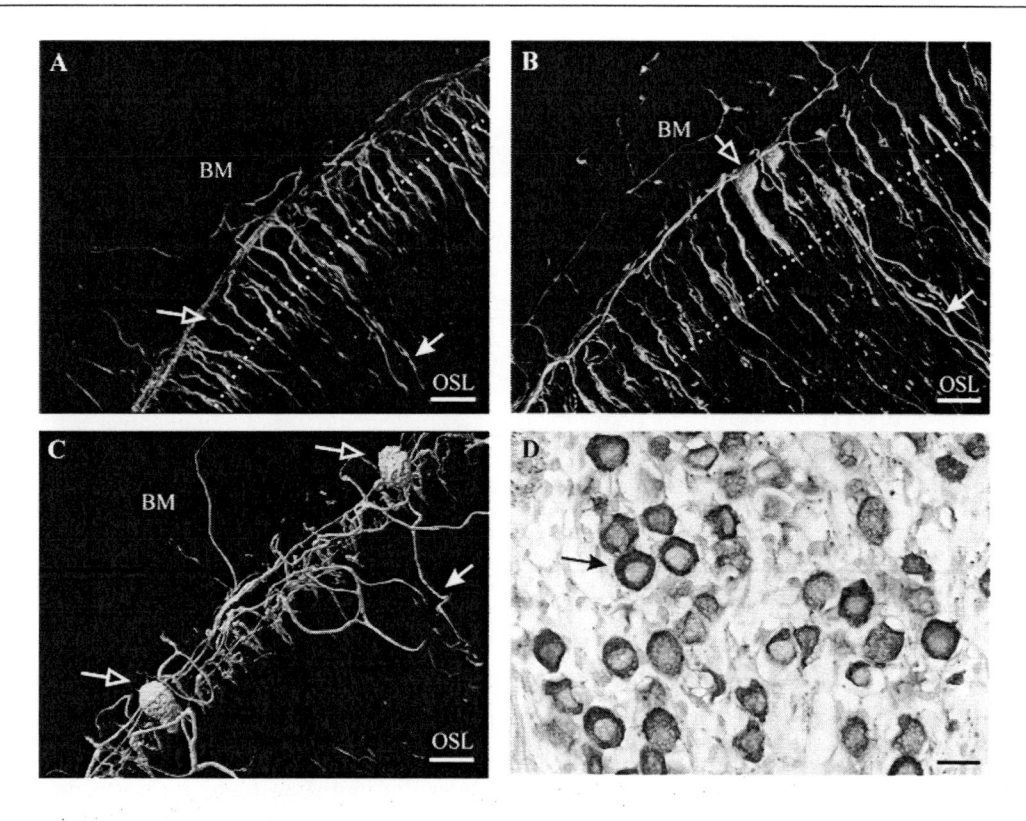

Figure 3. Whole mount preparations (A, B and C) showing a 'top view' (see figure 1) of the NF-H stained peripheral processes in the basal region of a short-term deafened cochlea (33 days). There was a marked reduction in the number of tunnel-crossing peripheral processes on the basilar membrane (BM) and in the habenula perforata region (dotted lines) very few processes could be observed (A - open arrow). Peripheral processes within the OSL, proximal to the habenula perforata, were stained with NF-H (filled arrows in A and B), indicative of their demyelination. Interestingly, accumulations of NF-H were observed in the distal ends of a number of degenerating peripheral processes (B and C – open arrow). A number of resprouting peripheral processes were also evident (C – filled arrow). A cochlear cross section (2μm) shows NF-H staining in the SGN cell bodies within Rosenthal's canal. Although there was a reduction in the number of SGNs after deafening the remaining SGNs were immunopositive for the NF-H antibody (D – arrow) (A-D scale bar 20μm).

Resprouting of Peripheral Processes and the Effect of Neurotrophin Administration

SGNs are incapable of postembryonic cellular mitosis, therefore it is important to determine whether damaged SGNs can be repaired. Resprouting of SGN peripheral processes has been reported following various forms of cochlear insult, for instance, ototoxic aminoglycoside antibiotics (Terayama et al., 1977; Terayama et al., 1979; Webster and Webster, 1982), acoustic trauma (Bohne and Harding, 1992; Strominger et al., 1995; Lawner et al., 1997) and pharmacological excitotoxicity (Puel et al., 1995). However, little is known about the processes that influence resprouting of auditory neurons *in vivo* and whether factors such as neurotrophins can facilitate this process. Neurotrophic factors, such as the neurotrophins brain-derived neurotrophic factor (BDNF) and neurotrophin-3 (NT-3) are well known to play important roles in the development and maintenance of the nervous system, including the auditory system. Research has shown that intracochlear infusion of various neurotrophic factors can rescue auditory neurons from the degeneration that typically occurs in deafness (Ernfors et al., 1996; Staecker et al., 1996; Miller et al., 1997; Gillespie et al., 2003; Gillespie et al., 2004; Richardson et al., 2005). We have carried out experiments to examine the ability of the peripheral processes in guinea pig cochleae to resprout following sensorineural hearing loss and to assess whether the administration of neurotrophins (BDNF and NT-3) can enhance or promote the resprouting process. A focus of this study was to examine the neurofilament profile of resprouting peripheral processes.

Guinea pigs (n=5) were deafened and five days later were implanted with a mini-osmotic pump (see Wise et al., 2005). The pump provided a constant slow-rate intracochlear infusion of the neurotrophins BDNF and NT-3 over a 28 day period. After 28 days of implantation the neurotrophin-treated and contralateral control (untreated) cochleae were harvested and processed for neurofilament immunohistochemistry. Comparisons were made between whole mounts generated for the neurotrophin-treated cochleae and the deafened (control) cochleae. Resprouting processes were readily identified by their abnormal projections, which were distinct to those observed in the normal cochlea. In the normal cochlea the peripheral processes of the auditory neurons have characteristic and uniform innervation patterns (Figure 2). In contrast, resprouting processes in the deafened cochleae were disorganized and followed rather torturous pathways (Figure 4). Resprouting peripheral processes were prominently stained with an NF-H antibody in both the untreated (deaf control) cochleae (Figure 4A-C) and in the neurotrophin-treated cochleae (Figure 4D). However, resprouting peripheral processes were observed over a significantly larger area in the basal region of the neurotrophin-treated cochleae (for details see Wise et al., 2005). Resprouting peripheral processes were observed to exit the habenula perforata and project in a basal or apical direction on the basilar membrane or on the inner sulcus (Figure 4A). They were also detected to loop within the OSL and to loop on the basilar membrane and enter openings of the habenula perforata that were located laterally from which they originated (Figure 4B). Most commonly, they were observed to project out of their habenula opening and towards the spiral limbus. Here, it was also apparent that some resprouting peripheral processes navigated their way through grooves between the inner sulcus cells in a honeycomb-like manner (Figure 4D).

Flattened structures with a similar morphology to growth cones were often observed at the distal ends of resprouting peripheral processes (Figure 4C). These structures were relatively high in neurofilament expression and exhibited fine filopodia-like projections that were also stained with the neurofilament antibody. The expression of GAP-43, which is a protein present in the growth cones of resprouting and developing neurons and which enables the formation of new connections, was not prominent within the growth cones at the time point examined in this study. In a small number of resprouting peripheral processes GAP-43 expression was observed more proximally (within the OSL) suggesting that this protein, whilst present, may not play a prominent role in the functioning of growth cones after a 33-day period of deafness. Conversely, the strong expression of NF-H in the growth cone structures implies that neurofilaments play an important role in resprouting and are likely to add structural support and possibly stabilise the resprouting processes in the extracellular environment.

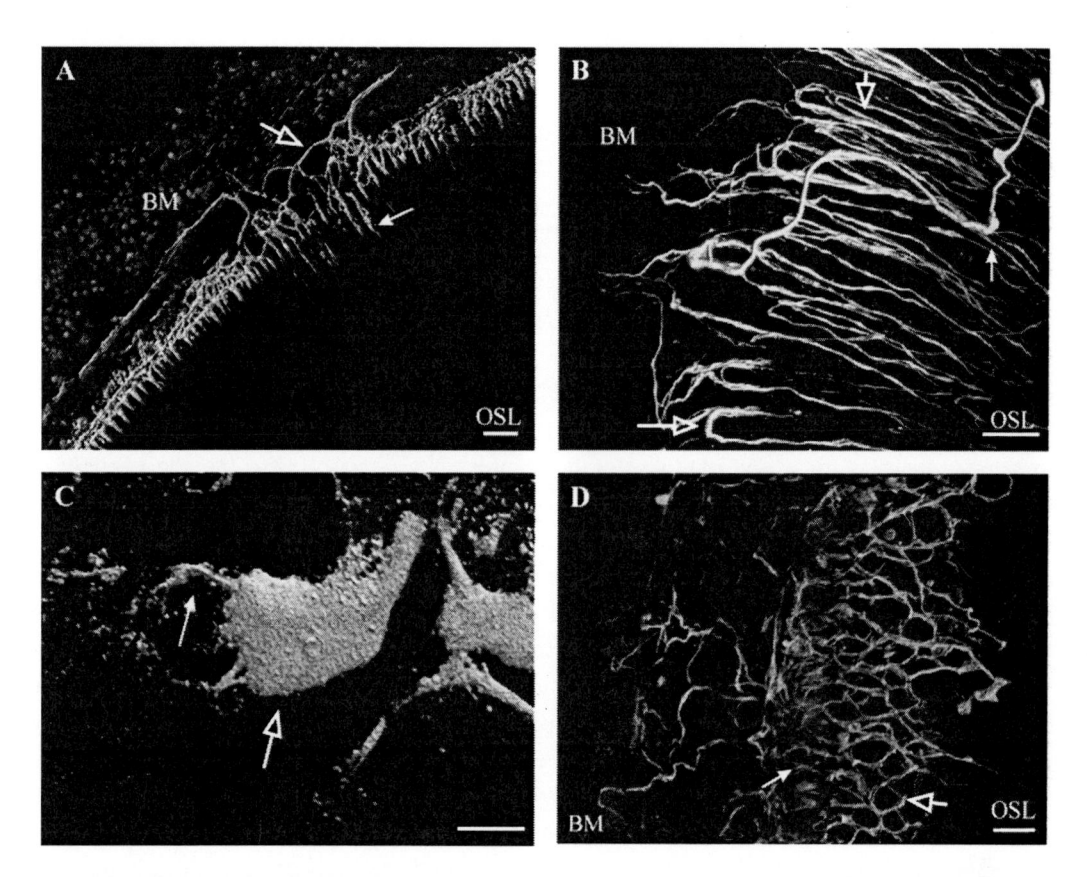

Figure 4. NF-H stained resprouting peripheral processes in the basal turn of the short-term deafened cochlea (33 days). A number of resprouting peripheral processes (A - open arrow) were seen projecting on the basilar membrane (BM), having originated from a region within the OSL that exhibited a greater degree of demyelination (filled arrow, scale bar 50µm). B Looping resprouting peripheral processes (open arrow – near bottom) were observed to exit the habenula perforata and re-enter a different habenular opening located distally and were also observed to reverse their direction within the OSL and extend towards the SGNs (open arrow near top). One resprouting process (filled arrow) was seen to exit the habenula opening and project back towards the spiral limbus, seemingly making points of contact with the wall of the inner sulcus (not stained), visible as bulges in the resprouting process (scale bar 20µm). C Some resprouting peripheral processes exhibited flattened structures at their distal end (open arrow) that appeared to be growth cones. Filopodia-like projections were observed at the leading edge (filled arrow, scale bar 5µm). D NF-H stained resprouting peripheral processes in the neurotrophin treated cochleae were observed to project from their habenula opening (filled arrows) and towards the spiral limbus. Some remaining peripheral processes were observed on the BM in the outer spiral bundle. The unmyelinated regions of the resprouting peripheral processes were observed in a honeycomb-like configuration (open arrow) as they course their way in the grooves between the inner sulcus cells. (scale bar 20µm). (Figure adapted from Wise et al., 2005).

It is well established that microtubules and actin filaments are involved in neuron outgrowth and stability. However, evidence is beginning to mount that neurofilaments may also play a key role in the regeneration process. In a study using knockout mice that lack NF-L, a 20-fold decrease in the levels of NF-M and NF-H was observed. Importantly, a delay and reduction in the extent of regeneration of myelinated axons after peripheral nerve crush was also observed, leading the authors to conclude that neurofilament proteins play a significant role in regeneration (Zhu et al., 1997). In experiments using a similar line of reasoning, the injection of anti-NF-M antibodies into *Xenopus laevis* embryos and cultures resulted in shorter axons that grew at an overall slower rate than in the normal experimental group. These results suggested that neurofilaments may stabilise the axon cytoskeleton during neurite extension and possibly inhibit the retraction of long axons (Lin and Szaro, 1995; Walker et al., 2001).

It is also possible that resprouting processes interact with various surface adhesion molecules which may, in turn, alter neurofilament function. For instance, it is known that ephrins B2 and B3, mediated by the EphA4 receptor, are expressed within the cochlea during development and are thought to be important neuronal signaling mechanisms used in neuronal pathfinding (van Heumen et al., 2000; Brors et al., 2003). Furthermore, interactions between the levels of immunoreactive neurofilament and neural cell adhesion molecules (N-CAM) have been reported in *in vitro* experiments, with higher neurofilament levels observed in sensory neurons grown on monolayers that express N-CAM isoforms, thereby suggesting that interactions between neurofilaments and cell surface adhesion molecules may be important in neurite outgrowth (Doherty et al., 1989). Although little is known about the expression of extracellular guidance and adhesion molecules within the adult cochlea following ototoxic deafening, it is possible that resprouting neurons can interact with these types of cues to influence neurofilament function.

The ultimate outcome of the resprouting auditory processes is not known. There is evidence from ototoxically deafened guinea pigs that resprouting peripheral processes only survive temporarily (Terayama et al., 1977; Terayama et al., 1979). However, others have reported resprouting peripheral processes 1-2 years following noise-induced damage to the organ of Corti (Strominger et al., 1995). The success or failure of axonal regeneration may be determined by intrinsic neuronal mechanisms that regulate the axonal cytoskeleton and also by factors that enable the transition from a migrating, target-seeking behaviour to that of a mature neuron that has formed a functional connection. A key factor that might influence the long-term success of resprouting peripheral processes is whether a neuronal target can be located and a functional connection restored. Therefore, in the long term, if a means to regenerate auditory hair cells within the deaf ear is discovered, the potential exists for establishment of such synaptic connections, and perhaps the restoration of a functional auditory system. Such a means of hair cell regeneration might not be that far from reality, since recent gene transfer experiments in guinea pigs have shown that expression of the *Math1* gene in supporting cells of the organ of Corti induced differentiation of the supporting cells into cells with the unique and distinctive morphology of hair cells (Kawamoto et al., 2003). Furthermore, there is evidence that the newly formed hair cells are also functioning in the same manner as normal hair cells since they caused regenerating neurons to alter their normal trajectory to synapse with them, presumably due to the release of neurotrophic factors, and also resulted in improved hearing thresholds (Kawamoto et al., 2003; Izumikawa et al., 2005).

Diameter of the Peripheral Processes in the Deafened Cochlea and the Effects of Neurotrophin Administration

Thin (2μm) cross sections were taken through the OSL of the osmium-stained cochlea (see Figure 1A) to enable measurement of the diameters of the myelinated peripheral processes in the deafened untreated and neurotrophin-treated cochleae. In addition to the pronounced demyelination and degeneration of the peripheral processes in the deafened cochleae, the remaining myelinated peripheral processes were smaller in diameter and often lacked a visible axoplasm (Figure 5A compared to B and C). The results indicated that the remaining myelinated peripheral processes were significantly smaller in diameter for both the short-term (33 day) and the long-term (61 day) deafened groups when compared to the control group (normal hearing) (Figure 5E and F, $p<0.001$ ANOVA). Decreases in the diameter of auditory neurons following sensorineural hearing loss have previously been reported and are a feature of auditory neuron degeneration (Ylikoski et al., 1974; Terayama et al., 1977; Terayama et al., 1979; Leake and Hradek, 1988; Shepherd and Javel, 1997; Araki et al., 1998). In a previous study investigating neural degeneration in the deafened cochlea, Ylikoski et al. (1974) observed a collapse in the myelin sheath with diminished electron density and the occurrence of nerve fibres with a darkened and shrunken axoplasm. The myelin sheath on the remaining peripheral processes within the OSL typically appeared normal but collapsed due to the diminished axoplasm. Changes to the neurofilament cytoarchitecture were thought to be involved, as there was often a reduction in the number of neurofilaments and microfilaments within these shrunken peripheral processes. It was thought that alteration of neurofilament functioning was likely to play an important role in this finding.

Previous studies have shown that the calibre of myelinated axons in large motor nerve fibres is determined by the number of neurofilaments (Hoffman et al., 1984). The radial growth of axons may correlate with neurofilament number, which in turn relies more on NF-L and NF-M expression, since both are more important to filament assembly than NF-H (Grant and Pant, 2000). The finding that NF-L null mutant mice exhibited reduced axonal diameters as well as reductions in NF-M and NF-H expression provided additional evidence of a role for neurofilament proteins in the maintenance of axon calibre (Zhu et al., 1997). Other studies using mice mutant for either NF-H or NF-M led to the conclusion that neither NF-H nor its phosphorylated tail is essential for determining neuronal calibre, and that NF-M may actually play a more important role in regulating axonal diameter (Liu et al., 2004). Furthermore, evidence suggests that Schwann cells interact closely with axons and that signals from Schwann cells can regulate the cytoskeletal proteins and axonal function. It has been shown that demyelination of axons decreases NF-M and NF-H sidearm phosphorylation leading to a decrease in neurofilament density and thus may contribute to the decrease in axon calibre observed (de Waegh et al., 1992).

Intracochlear administration of neurotrophins (BDNF and NT-3) following sensorineural hearing loss (short-term deafening) was effective in maintaining myelinated peripheral processes with diameters that were statistically the same as those in the normal cochleae

(Figure 5D,G). Importantly, in comparison to the deafened control cochleae (Figure 5B), the diameter of the peripheral processes were significantly larger following neurotrophin treatment (ANOVA, p<0.001), supporting the notion that neurotrophic factors may regulate neurofilament activity in order to maintain normal axon calibre.

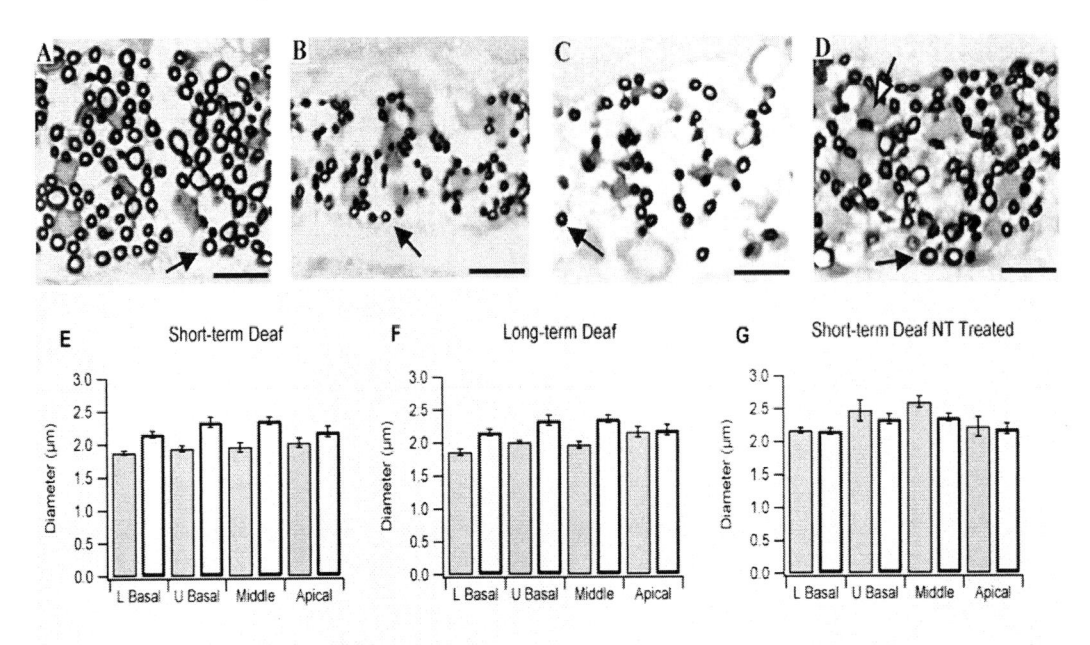

Figure 5. OSL cross sections (see figure 1 for details) from the basal region of a normal cochlea (A), a short-term deafened cochlea (33 days) (B), a long-term deafened cochlea (61 days) (C) and a short-term deafened, neurotrophin-treated cochlea (5 days deaf and 28 days neurotrophin delivery) (D). The osmium-stained myelinated peripheral processes (filled arrows) were smaller in diameter and fewer in number in the deafened cochleae compared to the normal cochlea (see E and F). Schwann cells were present in relatively high numbers in the neurotrophin treated cochleae (D, open arrow), (scale bar 10μm). The average (±SEM) of the fibre diameter is shown for the short-term deafened cochleae (E, grey bars, n=9), the long-term deafened cochleae (F, grey bars, n=13) and the short-term deafened, neurotrophin-treated cochleae (G, grey bars, n=5), each plotted with the fibre diameters from normal cochleae (white bars, n=10) plotted against cochlear position (L-lower, U-upper). The diameters of the peripheral processes in neurotrophin-treated cochleae were not statistically different to those observed in normal cochleae (G). (Figure adapted from Wise et al., 2005).

It has been suggested that the loss of neurotrophic signals from the target maybe involved in triggering this process (Gold et al., 1991). Treatment of transected sciatic nerve with nerve growth factor (NGF) was shown to enhance the diameter of dorsal root ganglion (DRG) axons as a result of an increase in the neurofilament content (Verge et al., 1990; Gold et al., 1991). Therefore, neurotrophic factors are thought to regulate neurofilament proteins, which is likely to be an important feature of the neuron survival promoting and regenerative capacities of these trophic agents.

The mechanism responsible for neurofilament function in maintaining the calibre of the peripheral processes may be closely linked to its state of phosphorylation (Liu et al., 2004). It has been reported that neurotrophic factors can regulate neurofilament phosphorylation. For example, treatment with BDNF or NT-3 stimulated the phosphorylation of NF-H in rat cortical neuron cultures (Tokuoka et al., 2000). In addition, NGF has been shown to induce neurite outgrowth from neuron-like pheochromocytoma (PC-12) cells which is accompanied by stimulation of NF-L and NF-M transcription (Traverse et al., 1992). This outcome was associated with a sustained increase in Erk1/2 activity, thereby implicating the MAP (mitogen-activated protein) kinase pathway as a mechanism for neurotrophin effects on axonal growth and regeneration.

The neuronal cytoskeleton is regulated by phosphorylation and dephosphorylation reactions that are mediated by numerous associated kinases and phosphatases (reviewed by Grant and Pant, 2000). The stress-activated protein kinase subgroup of MAP kinases are also important for NF-H tail domain phosphorylation, and are also downstream regulators of neurotrophins via the high affinity Trk receptors (Segal and Greenberg, 1996; Kaplan and Miller, 2000). Phosphorylation sites within the tail domains of NF-H and NF-M are also catalysed by Cdk5 (cyclin-dependent kinase 5), which is reported to be activated by BDNF to induce NF-H phosphorylation (Tokuoka et al., 2000). Since neurofilament phosphorylation has been correlated with neurofilament density and inter-filament spacing (Grant and Pant, 2000), this finding also supports the suggestion that neurotrophic factors may influence axonal calibre via neurofilament phosphorylation. In addition, phosphorylation of NF-M and NF-H protects neurofilaments from protease degradation (Goldstein et al., 1987; Pant, 1988; Perrone Capano et al., 2001), providing a potential mechanism for the greater extent of resprouting peripheral processes in neurotrophin-treated cochleae.

The most obvious functional implication of reduction in peripheral process diameter observed following deafening is a decrease in the conduction velocity of the associated neurons (see Miller et al., 2002). In the normal cochlea, Type I neurons are relatively large in diameter and typically have high spontaneous discharge rates with low activation thresholds to electrical stimulation (Liberman, 1982; Liberman and Oliver, 1984; Gleich and Wilson, 1993). In the deafened cochlea, however, ongoing structural changes to the auditory neurons may alter the dynamics of their electrical activation by devices such as the cochlear implant. The types of changes observed in response to neurotrophin administration could have significant ramifications for the functional properties of auditory neurons. Since neurotrophin administration was effective in maintaining greater numbers of SGNs with peripheral processes that were larger in diameter, electrical stimulation via a cochlear implant might be more effective and efficient. Supporting this suggestion are reports of functional changes after neurotrophin administration whereby the auditory brainstem response thresholds were significantly lowered in the cochleae that received neurotrophin treatment compared to the untreated cochlea (Shinohara et al., 2002; Shepherd et al., 2005). Therefore, any intervention that can maintain the functional integrity of the auditory neurons by preventing the degenerative process, or even promoting functional resprouting of the peripheral processes, could lead to a more effective cochlear implant.

Conclusion

Neurofilaments are important components of the neuronal cytoskeleton. They provide structural integrity and facilitate cell and organelle motility. Neurofilament proteins play a fundamental role in the development of the cochlea and the maintenance of neurons in the mature cochlea. Damage to the sensory hair cells in the cochlea can initiate the degeneration of the SGN peripheral processes that can ultimately lead to death of the SGNs. Demyelination, decreases in the diameter of the peripheral processes and the breakdown of the neurofilament cytoskeleton are prominent features of the degeneration process. However, these degenerative effects can be prevented, and to a certain extent reversed, by providing neurotrophic support via the intracochlear infusion of BDNF and NT-3. Neurotrophic factors have been shown to regulate neurofilament activity and phosphorylation, and therefore may play a key role in axonal growth and regeneration. Evidence supporting this suggestion was the finding that BDNF and NT-3 treatment resulted in greater numbers of SGNs with peripheral processes of normal diameter in addition to the enhanced resprouting of auditory peripheral processes. Resprouting peripheral processes were prominently stained with NF-H, and strong NF-H expression was also observed in growth cone-like structures at the distal ends of these processes. Therefore, it is likely that neurofilament proteins played an important role in providing structural support and stability through the course of regeneration. While there are conflicting reports as to the long-term survival of resprouting auditory processes, the success or failure of axonal regeneration may be determined by factors that regulate the axonal cytoskeleton, but also by the formation of a functional connection which appears necessary to stabilise regenerating neural processes.

The observation that the peripheral processes of the auditory neurons can regrow has clinical implications for cochlear implant patients. The possible regrowth of damaged peripheral processes towards a stimulating electrode array may result in an improved electro-neural interface and thus enhance the performance and efficacy of the cochlear implant. In the longer term, resprouting peripheral processes may be stimulated to grow towards a regenerated hair cell population, for restoration of a fully functional auditory system. An understanding of the mechanisms involved in nerve regeneration in the auditory system, in particular in relation to neurofilaments and neurotrophic factors, is a key factor towards realising these goals.

Acknowledgements

This research was funded by the Stavros S. Niarchos Foundation. We would like to thank Maria Clarke and Prudence Nielsen for assistance with the histology. Additional acknowledgements go to A/Prof. Stephen O'Leary for his contributions and Dr Ian Harper from Monash University Micro Imaging for his assistance with the confocal microscopy.

References

Angeletti R.H., Trojanowski J.Q., Carden M., Schlaepfer W.W. and Lee V.M. (1985). Domain structure of neurofilament subunits as revealed by monoclonal antibodies. *J Cell Biochem 27 (2)*: 181-187.

Anniko M., Thornell L.E. and Virtanen I. (1987). Cytoskeletal organization of the human inner ear. *Acta Otolaryngol Suppl 437*: 5-76.

Araki S., Kawano A., Seldon L., Shepherd R.K., Funasaka S. and Clark G.M. (1998). Effects of chronic electrical stimulation on spiral ganglion neuron survival and size in deafened kittens. *Laryngoscope 108 (5)*: 687-695.

Berglund A.M. and Ryugo D.K. (1991). Neurofilament antibodies and spiral ganglion neurons of the mammalian cochlea. *J Comp Neurol 306 (3)*: 393-408.

Bohne B.A. and Harding G.W. (1992). Neural regeneration in the noise-damaged chinchilla cochlea. *Laryngoscope 102 (6)*: 693-703.

Brors D., Bodmer D., Pak K., Aletsee C., Schafers M., Dazert S. and Ryan A.F. (2003). EphA4 provides repulsive signals to developing cochlear ganglion neurites mediated through ephrin-B2 and -B3. *J Comp Neurol 462 (1)*: 90-100.

Carden M.J., Trojanowski J.Q., Schlaepfer W.W. and Lee V.M. (1987). Two-stage expression of neurofilament polypeptides during rat neurogenesis with early establishment of adult phosphorylation patterns. *J Neurosci 7 (11)*: 3489-3504.

Cork L.C., Sternberger N.H., Sternberger L.A., Casanova M.F., Struble R.G. and Price D.L. (1986). Phosphorylated neurofilament antigens in neurofibrillary tangles in Alzheimer's disease. *J Neuropathol Exp Neurol 45 (1)*: 56-64.

Dau J. and Wenthold R.J. (1989). Immunocytochemical localization of neurofilament subunits in the spiral ganglion of normal and neomycin-treated guinea pigs. *Hear Res 42 (2-3)*: 253-263.

de Waegh S.M., Lee V.M. and Brady S.T. (1992). Local modulation of neurofilament phosphorylation, axonal caliber, and slow axonal transport by myelinating Schwann cells. *Cell 68 (3)*: 451-463.

Delisle M.B. and Carpenter S. (1984). Neurofibrillary axonal swellings and amyotrophic lateral sclerosis. *J Neurol Sci 63 (2)*: 241-250.

Doherty P., Barton C.H., Dickson G., Seaton P., Rowett L.H., Moore S.E., Gower H.J. and Walsh F.S. (1989). Neuronal process outgrowth of human sensory neurons on monolayers of cells transfected with cDNAs for five human N-CAM isoforms. *J Cell Biol 109 (2)*: 789-798.

Ernfors P., Duan M.L., ElShamy W.M. and Canlon B. (1996). Protection of auditory neurons from aminoglycoside toxicity by neurotrophin-3. *Nat Med 2 (4)*: 463-467.

Fechner F.P., Burgess B.J., Adams J.C., Liberman M.C. and Nadol J.B., Jr. (1998). Dense innervation of Deiters' and Hensen's cells persists after chronic deefferentation of guinea pig cochleas. *J Comp Neurol 400 (3)*: 299-309.

Finn J.T., Weil M., Archer F., Siman R., Srinivasan A. and Raff M.C. (2000). Evidence that Wallerian degeneration and localized axon degeneration induced by local neurotrophin deprivation do not involve caspases. *J Neurosci 20 (4)*: 1333-1341.

Forno L.S., Sternberger L.A., Sternberger N.H., Strefling A.M., Swanson K. and Eng L.F. (1986). Reaction of Lewy bodies with antibodies to phosphorylated and non-phosphorylated neurofilaments. *Neurosci Lett 64 (3)*: 253-258.

Gillespie L.N., Clark G.M., Bartlett P.F. and Marzella P.L. (2003). BDNF-induced survival of auditory neurons in vivo: Cessation of treatment leads to accelerated loss of survival effects. *J Neurosci Res 71 (6)*: 785-790.

Gillespie L.N., Clark G.M. and Marzella P.L. (2004). Delayed neurotrophin treatment supports auditory neuron survival in deaf guinea pigs. *Neuroreport 15 (7)*: 1121-1125.

Gleich O. and Wilson S. (1993). The diameters of guinea pig auditory nerve fibres: distribution and correlation with spontaneous rate. *Hear Res 71 (1-2)*: 69-79.

Gold B.G., Mobley W.C. and Matheson S.F. (1991). Regulation of axonal caliber, neurofilament content, and nuclear localization in mature sensory neurons by nerve growth factor. *J Neurosci 11 (4)*: 943-955.

Goldstein M.E., Sternberger N.H. and Sternberger L.A. (1987). Phosphorylation protects neurofilaments against proteolysis. *J Neuroimmunol 14 (2)*: 149-160.

Grant P. and Pant H.C. (2000). Neurofilament protein synthesis and phosphorylation. *J Neurocytol 29 (11-12)*: 843-872.

Guinan J.J., Jr., Warr W.B. and Norris B.E. (1983). Differential olivocochlear projections from lateral versus medial zones of the superior olivary complex. *J Comp Neurol 221 (3)*: 358-370.

Hafidi A. and Romand R. (1989). First appearance of type II neurons during ontogenesis in the spiral ganglion of the rat. An immunocytochemical study. *Brain Res Dev Brain Res 48 (1)*: 143-149.

Hafidi A., Despres G. and Romand R. (1990). Cochlear innervation in the developing rat: an immunocytochemical study of neurofilament and spectrin proteins. *J Comp Neurol 300 (2)*: 153-161.

Hafidi A. (1998). Peripherin-like immunoreactivity in type II spiral ganglion cell body and projections. *Brain Res 805 (1-2)*: 181-190.

Hall G.F. and Lee V.M. (1995). Neurofilament sidearm proteolysis is a prominent early effect of axotomy in lamprey giant central neurons. *J Comp Neurol 353 (1)*: 38-49.

Hardie N.A. and Shepherd R.K. (1999). Sensorineural hearing loss during development: morphological and physiological response of the cochlea and auditory brainstem. *Hearing Research 128 (1-2)*: 147-165.

Hasko J.A., Richardson G.P., Russell I.J. and Shaw G. (1990). Transient expression of neurofilament protein during hair cell development in the mouse cochlea. *Hear Res 45 (1-2)*: 63-73.

Hisanaga S., Kusubata M., Okumura E. and Kishimoto T. (1991). Phosphorylation of neurofilament H subunit at the tail domain by CDC2 kinase dissociates the association to microtubules. *J Biol Chem 266 (32)*: 21798-21803.

Hoffman P.N. and Lasek R.J. (1975). The slow component of axonal transport. Identification of major structural polypeptides of the axon and their generality among mammalian neurons. *J Cell Biol 66 (2)*: 351-366.

Hoffman P.N., Griffin J.W. and Price D.L. (1984). Control of axonal caliber by neurofilament transport. *J Cell Biol 99 (2)*: 705-714.

Izumikawa M., Minoda R., Kawamoto K., Abrashkin K.A., Swiderski D.L., Dolan D.F., Brough D.E. and Raphael Y. (2005). Auditory hair cell replacement and hearing improvement by Atoh1 gene therapy in deaf mammals. *Nat Med 11 (3)*: 271-276.

Kamakura K., Ishiura S., Sugita H. and Toyokura Y. (1983). Identification of Ca2+-activated neutral protease in the peripheral nerve and its effects on neurofilament degeneration. *J Neurochem 40 (4)*: 908-913.

Kaplan D.R. and Miller F.D. (2000). Neurotrophin signal transduction in the nervous system. *Curr Opin Neurobiol 10 (3)*: 381-391.

Kawamoto K., Ishimoto S., Minoda R., Brough D.E. and Raphael Y. (2003). Math1 gene transfer generates new cochlear hair cells in mature guinea pigs in vivo. *J Neurosci 23 (11)*: 4395-4400.

Kong J., Tung V.W., Aghajanian J. and Xu Z. (1998). Antagonistic roles of neurofilament subunits NF-H and NF-M against NF-L in shaping dendritic arborization in spinal motor neurons. *J Cell Biol 140 (5)*: 1167-1176.

Lariviere R.C. and Julien J.P. (2004). Functions of intermediate filaments in neuronal development and disease. *J Neurobiol 58 (1)*: 131-148.

Lawner B.E., Harding G.W. and Bohne B.A. (1997). Time course of nerve-fiber regeneration in the noise-damaged mammalian cochlea. *Int J Dev Neurosci 15 (4-5)*: 601-617.

Leake P.A. and Hradek G.T. (1988). Cochlear pathology of long term neomycin induced deafness in cats. *Hear Res 33 (1)*: 11-33.

Lefebvre P.P., Weber T., Rigo J.M., Staecker H., Moonen G. and Van De Water T.R. (1992). Peripheral and central target-derived trophic factor(s) effects on auditory neurons. *Hear Res 58 (2)*: 185-192.

Liberman M.C. (1982). Single-neuron labeling in the cat auditory nerve. *Science 216 (4551)*: 1239-1241.

Liberman M.C. and Oliver M.E. (1984). Morphometry of intracellularly labeled neurons of the auditory nerve: correlations with functional properties. *J Comp Neurol 223 (2)*: 163-176.

Lin W. and Szaro B.G. (1995). Neurofilaments help maintain normal morphologies and support elongation of neurites in Xenopus laevis cultured embryonic spinal cord neurons. *J Neurosci 15 (12)*: 8331-8344.

Liu Q., Xie F., Siedlak S.L., Nunomura A., Honda K., Moreira P.I., Zhua X., Smith M.A. and Perry G. (2004). Neurofilament proteins in neurodegenerative diseases. *Cell Mol Life Sci 61 (24)*: 3057-3075.

LoPachin R.M. and Lehning E.J. (1997). Mechanism of calcium entry during axon injury and degeneration. *Toxicol Appl Pharmacol 143 (2)*: 233-244.

Marques S.A., Taffarel M. and Blanco Martinez A.M. (2003). Participation of neurofilament proteins in axonal dark degeneration of rat's optic nerves. *Brain Res 969 (1-2)*: 1-13.

Martinez A.M. and Ribeiro L.C. (1998). Ultrastructural localization of calcium in peripheral nerve fibres undergoing Wallerian degeneration: an oxalate-pyroantimonate and X-ray microanalysis study. *J Submicrosc Cytol Pathol 30 (3)*: 451-458.

Miller C.C., Ackerley S., Brownlees J., Grierson A.J., Jacobsen N.J. and Thornhill P. (2002). Axonal transport of neurofilaments in normal and disease states. *Cell Mol Life Sci 59 (2)*: 323-330.

Miller J.M., Chi D.H., O'Keeffe L.J., Kruszka P., Raphael Y. and Altschuler R.A. (1997). Neurotrophins can enhance spiral ganglion cell survival after inner hair cell loss. *Int J Dev Neurosci 15 (4-5)*: 631-643.

Oesterle E.C., Lurie D.I. and Rubel E.W. (1997). Neurofilament proteins in avian auditory hair cells. *J Comp Neurol 379 (4)*: 603-616.

Pachter J.S. and Liem R.K. (1984). The differential appearance of neurofilament triplet polypeptides in the developing rat optic nerve. *Dev Biol 103 (1)*: 200-210.

Pant H.C. (1988). Dephosphorylation of neurofilament proteins enhances their susceptibility to degradation by calpain. *Biochem J 256 (2)*: 665-668.

Perrone Capano C., Pernas-Alonso R. and di Porzio U. (2001). Neurofilament homeostasis and motoneurone degeneration. *Bioessays 23 (1)*: 24-33.

Puel J.L., Saffiedine S., Gervais d'Aldin C., Eybalin M. and Pujol R. (1995). Synaptic regeneration and functional recovery after excitotoxic injury in the guinea pig cochlea. *C R Acad Sci III 318 (1)*: 67-75.

Richardson R.T., O'Leary S., Wise A., Hardman J. and Clark G. (2005). A single dose of neurotrophin-3 to the cochlea surrounds spiral ganglion neurons and provides trophic support. *Hear Res 204 (1-2)*: 37-47.

Romand R., Sobkowicz H., Emmerling M., Whitlon D. and Dahl D. (1990). Patterns of neurofilament stain in the spiral ganglion of the developing and adult mouse. *Hear Res 49 (1-3)*: 119-125.

Roy S., Coffee P., Smith G., Liem R.K., Brady S.T. and Black M.M. (2000). Neurofilaments are transported rapidly but intermittently in axons: implications for slow axonal transport. *J Neurosci 20 (18)*: 6849-6861.

Ruben R.J. (1967). Development of the inner ear of the mouse: a radioautographic study of terminal mitoses. *Acta Otolaryngol*: Suppl 220:221-244.

Schlaepfer W.W. (1987). Neurofilaments: structure, metabolism and implications in disease. *J Neuropathol Exp Neurol 46 (2)*: 117-129.

Scott D., Smith K.E., O'Brien B.J. and Angelides K.J. (1985). Characterization of mammalian neurofilament triplet proteins. Subunit stoichiometry and morphology of native and reconstituted filaments. *J Biol Chem 260 (19)*: 10736-10747.

Segal R.A. and Greenberg M.E. (1996). Intracellular signaling pathways activated by neurotrophic factors. *Annu Rev Neurosci 19*: 463-489.

Shepherd R.K. and Javel E. (1997). Electrical stimulation of the auditory nerve. I. Correlation of physiological responses with cochlear status. *Hear Res 108 (1-2)*: 112-144.

Shepherd R.K. and Hardie N.A. (2001). Deafness-induced changes in the auditory pathway: implications for cochlear implants. *Audiol Neurootol 6 (6)*: 305-318.

Shepherd R.K., Coco A., Epp S.B. and Crook J.M. (2005). Chronic depolarization enhances the trophic effects of brain-derived neurotrophic factor in rescuing auditory neurons following a sensorineural hearing loss. *J Comp Neurol 486 (2)*: 145-158.

Shinohara T., Bredberg G., Ulfendahl M., Pyykko I., Olivius N.P., Kaksonen R., Lindstrom B., Altschuler R. and Miller J.M. (2002). Neurotrophic factor intervention restores auditory function in deafened animals. *Proc Natl Acad Sci U S A 99 (3)*: 1657-1660.

Shintaku M., Ogura J. and Terashima A. (1996). "Neuritic conglomerates" in the cerebral cortex of a patient with Creutzfeld-Jakob disease. *Acta Neuropathol (Berl) 92 (3)*: 319-323.

Spoendlin H. (1972). Innervation densities of the cochlea. *Acta Otolaryngol 73 (2)*: 235-248.

Staecker H., Kopke R., Malgrange B., Lefebvre P. and Van de Water T.R. (1996). NT-3 and/or BDNF therapy prevents loss of auditory neurons following loss of hair cells. *Neuroreport 7 (4)*: 889-894.

Strominger R.N., Bohne B.A. and Harding G.W. (1995). Regenerated nerve fibers in the noise-damaged chinchilla cochlea are not efferent. *Hear Res 92 (1-2)*: 52-62.

Terada M., Yasuda H. and Kikkawa R. (1998). Delayed Wallerian degeneration and increased neurofilament phosphorylation in sciatic nerves of rats with streptozocin-induced diabetes. *J Neurol Sci 155 (1)*: 23-30.

Terayama Y., Kaneko Y., Kawamoto K. and Sakai N. (1977). Ultrastructural changes of the nerve elements following disruption of the organ of Corti. I. Nerve elements in the organ of Corti. *Acta Otolaryngol 83 (3-4)*: 291-302.

Terayama Y., Kaneko K., Tanaka K. and Kawamoto K. (1979). Ultrastructural changes of the nerve elements following disruption of the organ of Corti. II. Nerve elements outside the organ of Corti. *Acta Otolaryngol 88 (1-2)*: 27-36.

Tokuoka H., Saito T., Yorifuji H., Wei F., Kishimoto T. and Hisanaga S. (2000). Brain-derived neurotrophic factor-induced phosphorylation of neurofilament-H subunit in primary cultures of embryo rat cortical neurons. *J Cell Sci 113 (Pt 6)*: 1059-1068.

Traverse S., Gomez N., Paterson H., Marshall C. and Cohen P. (1992). Sustained activation of the mitogen-activated protein (MAP) kinase cascade may be required for differentiation of PC12 cells. Comparison of the effects of nerve growth factor and epidermal growth factor. *Biochem J 288 (Pt 2)*: 351-355.

van Heumen W.R., Claxton C. and Pickles J.O. (2000). Expression of EphA4 in developing inner ears of the mouse and guinea pig. *Hear Res 139 (1-2)*: 42-50.

Verge V.M., Tetzlaff W., Bisby M.A. and Richardson P.M. (1990). Influence of nerve growth factor on neurofilament gene expression in mature primary sensory neurons. *J Neurosci 10 (6)*: 2018-2025.

Walker K.L., Yoo H.K., Undamatla J. and Szaro B.G. (2001). Loss of neurofilaments alters axonal growth dynamics. *J Neurosci 21 (24)*: 9655-9666.

Waller A. (1850). Experiments on the section of glossopharyngeal and hypoglossal nerves of the frog and observations of the alternatives produced thereby in the structure of their primitive fibers. *Philos Trans R Soc Lond B Biol Sci 140*: 423.

Wang L., Ho C.L., Sun D., Liem R.K. and Brown A. (2000). Rapid movement of axonal neurofilaments interrupted by prolonged pauses. *Nat Cell Biol 2 (3)*: 137-141.

Webster D.B. and Webster M. (1982). Multipolar spiral ganglion neurons following organ of Corti loss. *Brain Res 244 (2)*: 356-359.

Webster M. and Webster D.B. (1981). Spiral ganglion neuron loss following organ of Corti loss: a quantitative study. *Brain Res 212 (1)*: 17-30.

Wise A.K., Richardson R., Hardman J., Clark G. and O'Leary S. (2005). Resprouting and survival of guinea pig cochlear neurons in response to the administration of the neurotrophins brain-derived neurotrophic factor and neurotrophin-3. *J Comp Neurol 487 (2)*: 147-165.

Ylikoski J., Wersall J. and Bjorkroth B. (1974). Degeneration of neural elements in the cochlea of the guinea-pig after damage to the organ of corti by ototoxic antibiotics. *Acta Otolaryngol Suppl 326*: 23-41.

Zhu Q., Couillard-Despres S. and Julien J.P. (1997). Delayed maturation of regenerating myelinated axons in mice lacking neurofilaments. *Exp Neurol 148 (1)*: 299-316.

In: New Research on Neurofilament Proteins
Editor: Roland K. Arlen, pp. 25-51

ISBN: 1-60021-396-0
© 2007 Nova Science Publishers, Inc.

Chapter II

Neurofilament Proteins in Brain Diseases

Olivier Braissant[*]
Clinical Chemistry Laboratory, University Hospital of Lausanne, Switzerland

Abstract

Neurofilaments are the main components of intermediate filaments in neurons, and are expressed under three different subunit proteins, NFL, NFM and NFH. Neurofilaments act with microtubules and microfilaments to form and maintain the neuronal structure and cell shape. Phosphorylation is the main post-translational modification of neurofilaments, which influences their polymerization and depolymerization, and is responsible for their correct assembly, transport, organization and function in the neuronal process. In particular, phosphorylation is essential for the repulsion of the neurofilament polymers in axons, which determines the axonal diameter and the velocity of electrical conduction. The phosphorylation state of neurofilaments is regulated in a complex manner, including interactions with the neighbouring glial cells.

Abnormal expression, accumulation or post-translational modifications of neurofilament proteins are found in an increasing number of described neurological diseases, such as amyotrophic lateral sclerosis, Parkinson's, Alzheimer's and Charcot-Marie-Tooth diseases, or giant axonal neuropathy. Some of these diseases are associated with mutations discovered in the neurofilament genes. Recently, altered expression and phosphorylation states of neurofilament proteins have also been shown in metabolic diseases affecting the central nervous system either during development or in adulthood, such as hepatic encephalopathy due to hyperammonemia, methylmalonic and propionic acidemias, and diabetic neuropathy. Finally, accumulation of neurofilament proteins in

---

[*] Correspondence to: Olivier Braissant. Clinical Chemistry Laboratory, University Hospital of Lausanne, CH-1011 Lausanne, Switzerland; Tél: (+41.21) 314.41.52; Fax: (+41.21) 314.35.46; e-mail: Olivier.Braissant@chuv.ch

the cerebrospinal fluid has been described as discriminating marker for patients with multiple sclerosis, and as predictor of long-term outcome after cardiac arrest. This review will focus on the most recent investigations on neurofilament proteins in neurodegenerative, neurodevelopmental and metabolic diseases, as well as on the use of neurofilaments as markers of diseases.

Keywords: Neurofilaments, phosphorylation, neurodegenerative diseases, metabolic diseases, neurodevelopmental diseases, axon.

Introduction

Three types of filament proteins compose the cellular cytoskeleton: microtubules (Ø: ~25 nm), microfilaments (Ø: ~7 nm) and intermediate filaments (IFs; Ø: ~10 nm). Microtubules are essentially made of tubulin, and are involved in maintaining cell shape, in mitosis (formation of spindle fibers) and in the mouvement of organelles or vesicles. Actin is the main component of microfilaments, which are responsible for cell movements, muscular contraction, cytokinesis, mechanical strength, and, more specifically in CNS, axonal outgrowth and synaptic plasticity. Depending of the cell identity, a greater variety of proteins are found in IFs, which are prominent in cells that must withstand important mechanical stress, and are classified in five different types. The most important IFs in neurons are neurofilaments (NFs), which belong to type IV IFs and are exclusively neuronal. NFs establish an extremely stable tubular system of the neuronal cytoskeleton, having a 10 nm diameter. While NFs have been identified as structures since more than 100 years with the discovery of the silver staining technique, their precise roles in neuronal cytoskeleton have remained elusive until recently.

NFs are heteropolymers made of 3 different subunits: light (NFL), medium (NFM) and heavy (NFH) chain neurofilaments (Figure 1). These subunits assemble in a filamentous structure composing the main part of the axonal cytoskeleton. NFs interact with neighbouring cellular structures or other elements of the cytoskeleton through side arms protruding ouside of their filamentous structure. Their assembly in heteropolymers, as well as their interactions with neighbouring cellular structures, are regulated by post-translational modifications, from which the most important is phosphorylation, occuring in their head and side arms domains (Figure 1). Part of these post-translational modifications of NFs are regulated by glial cells in axonal vicinity. NFs participate to the rigidity of the axon, to its tensile strength, and to the regulation of axonal calibre. In that sense, NFs are essential to the formation and maintenance of the neuronal cell shape, and particularly of the axon, a structure with a diameter of 1 to 25 µm extending sometimes 100'000 times farther (1 m or more) than the neuronal cell body (10 to 50 µm in diameter). NFs also participate to the transport guidance of organelles and particles along the axon.

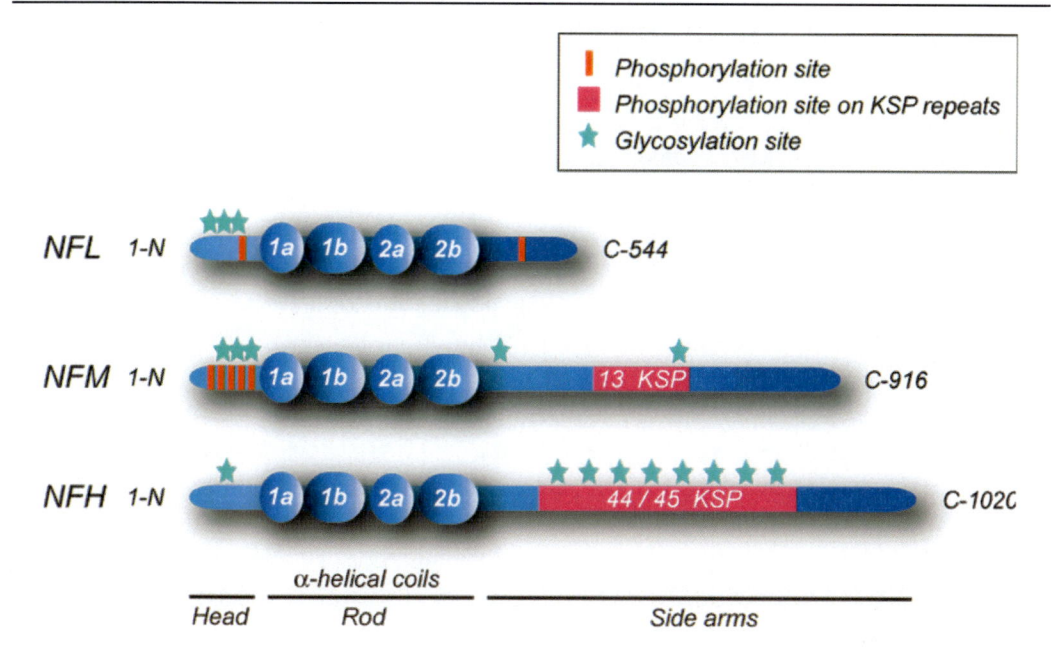

Figure 1. Schematic representation of human NFL, NFM and NFH proteins. Head, rod (α-helical coils) and side arms domains are indicated, as well as phosphorylation (including on KSP repeats) and glycosylation sites.

These last years, an increasing list of human brain diseases have been associated with NFs proteins. NFs proteins per se can be altered, either by mutations in their genes, or by alteration of their post-translational modifications, and particularly their phosphorylation state. The abnormal accumulation of neurofilaments have been observed in many neurodegenerative diseases, including Parkinson's disease (PD), amyotrophic lateral sclerosis (ALS), Alzheimer's disease (AD) or Charcot-Marie-Tooth (CMT) disease. More recently, altered expression and phosphorylation states of NFs have also been shown in metabolic diseases affecting the central nervous system either during development or in adulthood, such as hepatic encephalopathy due to hyperammonemia, methylmalonic and propionic acidemias, and diabetic neuropathy. Finally, the extracellular release of NFs proteins, due to axonal mechanical break-down or damage, and their accumulation in the cerebrospinal fluid can be followed as discriminating markers for patients with multiple sclerosis, and as predictor of long-term outcome after cardiac arrest.

This review will discuss NFs proteins expression and assembly in filamentous tubular structures, as well as their post-translational modifications. Focus will be made on the most recent NFs investigations in neurodegenerative, neurodevelopmental and metabolic diseases, and on the use of NFs as markers of diseases.

Neurofilament Proteins

NFs, as peripherin, α-internexin and nestin, belong to type IV IFs, with which they share common sequence structures. Three NF subunits contribute to the assembly of neurofilaments: Light (NFL), medium (NFM) and heavy (NFH) chain NFs (Figure 1). Human

NFL is encoded by the *NEFL* gene located on chromosome 8 (8p21) and consists of 544 amino acids. Human NFM is encoded by the *NEFM* gene also located on chromosome 8 (8p21) and consists of 916 amino acids. Human NFH is encoded by the *NEFH* gene located on chromosome 22 (22q12.2) and consists of 1020 amino acids. NFL, NFM and NFH have a molecular weight of 60, 100 and 110 kDa respectively, calculated on their amino acid sequence; however, due to important posttranslational modifications (i.e. phophorylation and glycosylation), NFL, NFM and NFH exhibit higher molecular weight on SDS-PAGE: 68 kDA, 160 kDa and 205 kDa respectively (for reviews, see: Lee and Cleveland, 1996; Parry and Steinert, 1999b; Al-Chalabi and Miller, 2003; Liu et al., 2004; Lariviere and Julien, 2004).

NFs are exclusively expressed by neurons. IFs, including NFs, are expressed differentially during CNS development and maturation. Undifferentiated brain cells express the type III IF protein vimentin (Bignami et al., 1982; Cochard and Paulin, 1984), while later neuroblasts express nestin, α-internexin, and peripherin (Portier et al., 1983; Lendahl et al., 1990; Kaplan et al., 1990). The neuronal differentiation induces the expression of NFs (Shaw and Weber, 1982; Carden et al., 1987; Nixon and Shea, 1992). NFL appears first at the start of neuronal differentiation, overlapping with α-internexin and peripherin expression (Willard and Simon, 1983; Carden et al., 1987). NFM follows NFL shortly after, when neurite elongation starts, *NEFL* and *NEFM* genes being located on the same chromosome and regulated in coordination. NFH appears later during axonal maturation (Willard and Simon, 1983; Carden et al., 1987).

NFs, as all IF proteins, share a common structure. In the centre of the protein, a rod domain of approximately 310 amino acids forms highly conserved α-helical motifs (regions 1a, 1b, 2a and 2b, Figure 1). Every seventh residue in this central rod domain is hydrophobic, facilitating the formation of α-helical coiled-coil parallel homo- or heterodimers (see below). The central rod domain is flanked by less conserved aminoterminal globular head and carboxyterminal side-arm tail. Head and tail confer their functional specificities to the different IF proteins: whilst the central rod domain is mainly responsible for NF assembly, head and tail interact with the environment of NFs (e.g. protein-protein interactions or axonal diameter) (Heins et al., 1993). The head domain also contributes to NF assembly (Gill et al., 1990). NFs are obligate heteropolymers in vivo, with NFL being required to form proper heteropolymers with either NFM or NFH (Lee et al., 1993; Ching and Liem, 1993). The dimer is formed by the head to tail coiled apposition of two NF proteins (NFL and either NFM or NFH) by their central rod domain. Two NF dimers assemble then in an half-staggered antiparallel NF tetramer (Cohlberg et al., 1995). The final 10 nm filament of NFs is formed by the lateral and longitudinal helical association of eight NF tetramers (Heins and Aebi, 1994; Fuchs and Weber, 1994; Fuchs and Cleveland, 1998; Parry and Steinert, 1999a; Herrmann and Aebi, 2000). The other IF proteins α-internexin and peripherin may also co-assemble, as homodimers however, with the NF heterodimers, especially during development (α-internexin, peripherin) and in restricted sets of mature neurons (peripherin) (Kaplan et al., 1990; Fliegner et al., 1994; Beaulieu et al., 1999). During neuronal differentiation (i.e.: neurite formation, axonal growth and maturation), the nature of the NF fibers changes, starting with heterodimers NFL-NFM only, followed, once NFH starts its expression, by NF fibers constituted of NFL-NFM and NFL-NFH heterodimers (Carden et al., 1987). Along time, a specific NF tetrameric unit can be replaced by another, explaining the differential stoichiometry observed in the NF fibers from development to mature CNS, influencing also axonal structure and functions.

NFL is essential for the precise NF assembly and for the maintenance of axonal calibre (Zhu et al., 1997). NFM participates in cross-bridges between NF fibers, stabilizes the NF filament network, participates in neurite longitudinal extension, and influences the axonal radial growth (Elder et al., 1998a; Jacomy et al., 1999; Elder et al., 1999a; Elder et al., 1999b). NFH also contributes to cross-bridges between NF fibers and may interact with microtubules, microfilaments and other cytoskeletal elements (Elder et al., 1998b; Jacomy et al., 1999; Elder et al., 1999b). In contrast to NFM, NFH does not seem to influence the axonal radial growth (Rao et al., 2002b).

NFs, after synthesis in the neuronal cell body, are then rapidly transported into the axons. Until recently, it was not clear whether NFs were transported into the axon as polymeric structures (« polymer hypothesis »), or as individual subunits (« subunit hypothesis ») (Baas, 1997; Hirokawa, 1997; Nixon, 1998). Radioisotopic pulse labeling studies argued for the polymeric hypothesis with NFs moving slowly in axons at an average rate of 0.2 to 1 mm/day, a much slower speed than any known axonal transport (Xu and Tung, 2000). On the other side, photobleaching experiments with fluorescence-tagged NFs argued for the subunit hypothesis, with the bleached axonal segment remaining stationary and slowly recovering its fluorescence (Okabe et al., 1993). The solution to this controversy came from recent works using live cell imaging and GFP-tagged NFs, that showed a fast transport of NF polymers (bursts of average speed of 1 to 2 mm/s) interrupted by prolonged pauses (Roy et al., 2000; Wang et al., 2000). As these fast bursts of NFs transport can be bidirectional, and due to the high proportion of paused NF fibers (> 90%), the resulting overall NF transport appears slow. NFs seem to use the conventional kinesin and dynein motor system (Shah et al., 2000; Yabe et al., 2000), and appear to dissociate from these motor systems after phosphorylation (Yabe et al., 1999). NFs are also translocated in dendrites of specific types of neurons, and seem required for the proper dendritic arborization of large motor neurons (Kong et al., 1998; Zhang et al., 2002).

Two major modifications are added post-translationally on NFs: phosphorylation and glycosylation. These modifications are dynamic and thought to regulate assembly, transport, structure and functions of NFs.

Various phosphorylation sites have been identified in the head (N-terminal) and tail (C-terminal) regions of NFs.

The head region of NFL and NFM can be phosphorylated at different positions (Figure 1) by protein kinases A, C and N (Sihag and Nixon, 1989; Sihag and Nixon, 1991; Hisanaga et al., 1994; Mukai et al., 1996; Cleverley et al., 1998; Nakamura et al., 2000). The phosphorylation of the NFL and NFM head region occurs rapidly after protein synthesis in the neuronal cell body, and inhibits the NF filament assembly in perikaria (Gibb et al., 1996; Gibb et al., 1998; Ching and Liem, 1999). This phosphorylation is transient, and the dephosphorylation of the NFL and NFM head region is a prerequisite for the axonal NF assembly in filaments (Gibb et al., 1998). Moreover, the transient phosphorylation of the head region of NFM also inhibits the phosphorylation of its C-terminal tail region (Zheng et al., 2003). Thus, before NF translocation in the axons, the phosphorylation of the head region of NFL and NFM protects neurons from a pathological accumulation of NF aggregates in their cell bodies.

Upon entry of NFs into the axon, the C-terminal side-arm domain of NFM and NFH, as well as the short C-terminal region of NFL, become phosphorylated. In particular, NFM and NFH are phosphorylated on Lys-Ser-Pro (KSP) repeat domains (Figure 1). In humans, NFM has 13 KSP repeats, while NFH exists with two polymorphic forms of either 44 or 45 KSP repeats (Figlewicz et al., 1993). Most of the serine residues of the KSP repeats can be phosphorylated, meaning that each mole of NFM and NFH can contain about 10 and 50 moles of phosphate, respectively (Julien and Mushynski, 1982; Grant and Pant, 2000). In axons, more than 99% of assembled NFM and NFH proteins are phosphorylated on their KSP repeats, in particular in myelinated internodal regions, while this proportion is much weaker in cell bodies, dendrites and nodes of Ranvier (de Waegh et al., 1992; Hsieh et al., 1994). Unphosphorylated NFs represent only ~1% of total NFs in the neurons. In the axon, NFs are phosphorylated in a proximal to distal gradient (Sternberger and Sternberger, 1983; Pant and Veeranna, 1995). The C-terminal region of NFL is phosphorylated by caseine kinase II (Nakamura et al., 1999), while the kinases that phosphorylate NFM and NFH KSP repeats in their C-terminal tail domains include GSK-3α/β, cdk5/p35, ERK1/2 and JNK1/3 (Guan et al., 1991; Giasson and Mushynski, 1996; Sun et al., 1996; Li et al., 2001).

In axons, the phosphorylation of multiple KSP repeats increases the negative charge of NFM and NFH, resulting in side-arm formation of their C-terminal tail and increased inter-neurofilament spacing (Nixon et al., 1994). This allows the radial axonal growth (i.e. regulation of axonal caliber), which increases axonal conduction velocity (de Waegh et al., 1992; Yin et al., 1998). The C-terminal phosphorylation of NFs also slows down their transport rate in axons, and mediate interactions with other cytoskeleton proteins, in particular microtubules (Hisanaga et al., 1991; Yabe et al., 2001; Shea et al., 2003). The phosphorylation of NFM seems preferentially responsible for the radial axonal growth, while the phosphorylation of NFH acts on the NF transport rate and their interactions with other proteins (Lewis and Nixon, 1988; Rao et al., 1998; Rao et al., 2003). The myelination of axons, both by Schwann cells in peripheral nerves and by oligodendrocytes in CNS, promotes the phosphorylation of NFM and NFH C-terminal tail, thus promoting the radial growth of myelinated axons and increasing their conduction velocity (de Waegh et al., 1992; Sanchez et al., 1996; Yin et al., 1998; Sanchez et al., 2000).

NFL, NFM and NFH are also post-translationally glycosylated by addtion of O-linked N-acetylglucosamine moieties on serine and threonine residues located in their head regions (NFL, NFM and NFH) as well as in their KSP repeat carboxyterminal region (NFM and NFH) (Figure 1) (Dong et al., 1993; Dong et al., 1996). The proximity of the O-GlcNAcylation and phosphorylation sites in the NF head domain suggest that competition between the two modes of post-translational modifications regulates NF assembly (Gill et al., 1990; Wong and Cleveland, 1990; Chin et al., 1991; Dong et al., 1993). On the other hand, in the nodes of Ranvier where NFs are more closely packed than in the internode axonal segments, O-GlcNAcylation probably replaces phosphorylation in the carboxyterminal KSP repeat region of NFM and NFH, rendering interactions between NFs more attractive than repulsive.

Therefore, phosphorylation / dephosphorylation and glycosylation / deglycosylation of NFs (by kinase / phosphatase and O-GlcNAc transferase / N-acetyl-β-D-glucosaminidase respectively) contributes to the assembly, structure and functions of NFs (Dong et al., 1993; Nixon, 1993; Xu et al., 1994; Dong et al., 1996).

Many neurons extend very long axons, up to 1 m in humans. To maintain the integrity and functions of these axons, some of their structural proteins, including those of the axonal cytoskeleton, have long lifetimes. For NFs in the human sciatic nerve, this average lifetime was estimated to 1 to 2 years (Lee and Cleveland, 1996). This very high stability of NFs is thought to be due, at least in part, to their phosphorylation which protects them from protease degradation (Goldstein et al., 1987; Pant, 1988). In physiological conditions, NF degradation only occurs in the axon terminus (presynaptic compartment), where NFs are dephosphorylated by protein phosphatase 2A (PP2A) (Gong et al., 2003), and then digested by calmodulin, a Ca^{++}-dependent protease (Maxwell et al., 1997).

Apart from their major role in regulating axonal caliber in function of their state of phosphorylation, NFs have been demonstrated or are postulated to have other functions in the axon. While gene knockout experiments demonstrated that NFs are not essential for axonal elongation, they nevertheless might facilitate it by stabilization of cytoskeletal elements and inhibition of axonal retraction (Zhu et al., 1997; Elder et al., 1998a; Elder et al., 1998b; Elder et al., 1999a). NFs participate, together with microtubules and microfilaments, to the axonal structural integrity, to the neuronal shape as well as to the axonal mechanisms of transport. They do so by direct or indirect interactions with microtubules (Hisanaga et al., 1991) or motor proteins like dynein, kinesin and myosin Va (Yabe et al., 1999; Shah et al., 2000; Yabe et al., 2000; Rao et al., 2002a), or with other crosslinking proteins like dystonin (Yang et al., 1999; Chen et al., 2000). NFM has been shown to interact with the D(1) dopamine receptor in subsets of neurons (Kim et al., 2002). Finally, of peculiar importance for the neuronal and axonal long term stability, NFs seem to protect axons from toxic components, by sequestrating for example Cdk5/p25 complexes which induce apoptosis (Nguyen et al., 2001), or by coupling of carbonyl groups issued of the oxidative stress on the lysine residues of KSP repeats (Wataya et al., 2002).

Neurofilament Proteins in Brain Diseases

As discussed above, the tight regulation of NF subunits expression, post-translational modifications, stoichiometry between NFL, NFM and NFH, and NF axonal transport, allows the correct assembly of NF filaments. This in turn contributes to the normal axonal growth, maturation, and stability along time. Any dysregulation of these precise mechanisms of NF regulations is susceptible to induce severe pathological consequences on neurons. In particular, the hallmark of numerous human neurological diseases is the abnormal accumulation of NFs in neuronal perikarya (for recent reviews, see Al-Chalabi and Miller, 2003; Liu et al., 2004; Lariviere and Julien, 2004; Petzold, 2005), which alters axonal growth, mechanisms of particles and organelles transportation, stability, and dynamic of interactions between NFs and other axonal proteins (Herrmann and Griffin, 2002). For a long time, it was admitted that NF abnormalities in human neurological disorders were secondary to neuronal dysfunctions. Recent studies demonstrate however that dysregulations of NFs themselves can be the cause of these pathologies. The second part of this review will focus on NF dysregulations in neurodegenerative, neurodevelopmental and metabolic diseases of central and peripheral nervous systems, as well as on the use of NFs as markers of specific diseases.

NFs in Neurodegenerative Diseases

Amyotrophic Lateral Sclerosis (ALS)

ALS is a progressive neurodegenerative disease affecting motor neurons in the brain and spinal cord, with a typical onset between 40 and 60 years of age. ALS patients usually die wihin 5 years after ALS diagnosis, due to motor neurons death and loss of function of the relative innervated muscles, and progressive partial or total paralysis. Most of the cognitive functions in ALS patients remain preserved. ALS is a heterogeneous syndrom, in which the neuropathological hallmark is an abnormal aggregation of NFs in the degenerating motor neurons (Manetto et al., 1988; Munoz et al., 1988). 5-10% of ALS cases are familial (autosomal dominant), while all the remaining cases are sporadic. 1-2% of all ALS cases (20-25% of familial ALS cases) are due to mutations in the Cu/Zn superoxide dismutase (SOD1) gene (Andersen, 2006), while the basis of the remaining ALS cases is still not known with precision. Mutations in SOD1 are thought to be linked to abnormal accumulation of NFs in ALS (Rouleau et al., 1996). Due to the abnormal accumulation and aggregation of hyperphosphorylated NFs in the ALS degenerating neurons, mutations in the NF genes have also been sought for a long time as good causative candidates for ALS. Indeed, different mutations have been found in NFs, in association with ALS (Figures 2,3,4). Codon deletions and insertions have been identified in the KSP regions of NFH in association with few sporadic cases of ALS (Figure 4) (Figlewicz et al., 1994; Tomkins et al., 1998; Al-Chalabi et al., 1999). More recently, missense mutations have also been found in the head and rod domains of NFH in other ALS cases (Garcia et al., 2006) (Figure 4). In association with ALS, the same group also identified recently a deletion in the tail domain of NFL (Figure 2), as well as missense mutations in the head, rod and tail domains of NFM (Figure 3) (Garcia et al., 2006). However, none of the mutations found in NF genes have been clearly identified as causative agent of ALS, nor linked to the familial dominantly inherited ALS (Al-Chalabi and Miller, 2003; Garcia et al., 2006), and it is thought now that these mutations in NF genes have to be considered as risk factors for sporadic ALS. However, the alteration of NF homeostasis seems to be an important part of the pathogenesis of ALS (Figures 2,3,4). As shown with mutant SOD1 transgenic models of ALS (Nguyen et al., 2001), the deregulation of specific NF kinase pathways (e.g: cdk5/p35) might cause the aberrant hyperphosphorylation of NFH and NFM side arms. This in turn might slow the axonal transport of NFs, which accumulate in neuronal perikarya (Williamson and Cleveland, 1999). The abnormal accumulation of NFs in the ALS degenerating neurons has also been associated with a significative decrease of NFL mRNA, which could increase the imbalance between NF subunits and precipitate further the neuronal degeneration (Bergeron et al., 1994;

Wong et al., 2000). This decrease in NFL mRNA seems due to the direct binding of mutant SOD1 to NFL mRNA, which destabilizes it (Ge et al., 2005). Interestingly, the two main posttranslational modifications of NFs, i.e. phosphorylation and glycosylation, might be conversely deregulated in ALS, as O-glycosylation of the C-terminal tail domain of NFM is decreased, while its phosphorylation is increased, in a transgenic rat model of ALS (Ludemann et al., 2005).

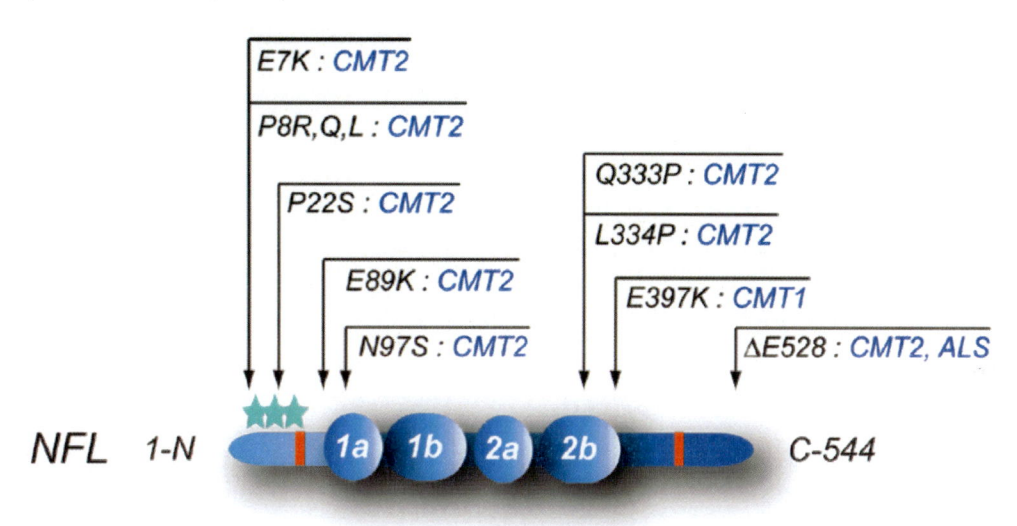

Figure 2. Schematic representation of NFL alterations in various brain diseases. Mutations identified in association with diseases are indicated above the NFL scheme, while the disease effects on NFL are indicated below the NFL scheme. AD: Alzheimer's disease; ALS: amyotrophic lateral sclerosis; CMT1, CMT2: Charcot-Marie-Tooth disease; PD: Parkinson's disease; Δ: deletion.

Charcot-Marie-Tooth Disease (CMT)

CMT is the most common inherited neurological disorder of the peripheral nervous system, affecting 1-4:10'000 individuals. CMT clinical phenotype is characterized by the progressive degeneration of motor and sensory neurons in the distal part of the limbs, leading to the slow loss of normal use of feet, legs, arms and hands (Skre, 1974; Reilly, 2000). CMT neuropathies are heterogeneous in the genes involved and, based on electrophysiological criteria, are classified in CMT1, a primary demyelinating form with reduced nerve conduction velocities, and CMT2, a primary axonal loss form. Some forms of CMT with overlapping characteristics between CMT1 and CMT2 have been classified as intermediate CMT. CMT is generally inherited with an autosomal dominant pattern. Recently, different missense mutations and one amino acid deletion have been identified in the *NEFL* gene (coding NFL) in several families in association with CMT (Figure 2) (Mersiyanova et al., 2000; De et al., 2001; Georgiou et al., 2002; Yoshihara et al., 2002; Jordanova et al., 2003; Choi et al., 2004; Zuchner et al., 2004). All these mutations are associated with the primary axonal loss form CMT2, with the exception of Glu397Lys being associated with the demyelinating form CMT1. These mutations in NFL are thought to disrupt NF assembly and axonal transport, as well as to alter NFL post-translational modifications. Other forms of CMT (CMT1) are caused by mutations in genes primarily expressed in Schwann cells and involved in myelin formation. These mutations lead to alterations in myelination, which in turn alter NFL, NFM and NFH phosphorylation states (Watson et al., 1994). The disruption of NF assembly and the alteration of NF phosphorylation states are thought to contribute, at least in part, to the CMT disease mechanisms leading to axonal degeneration.

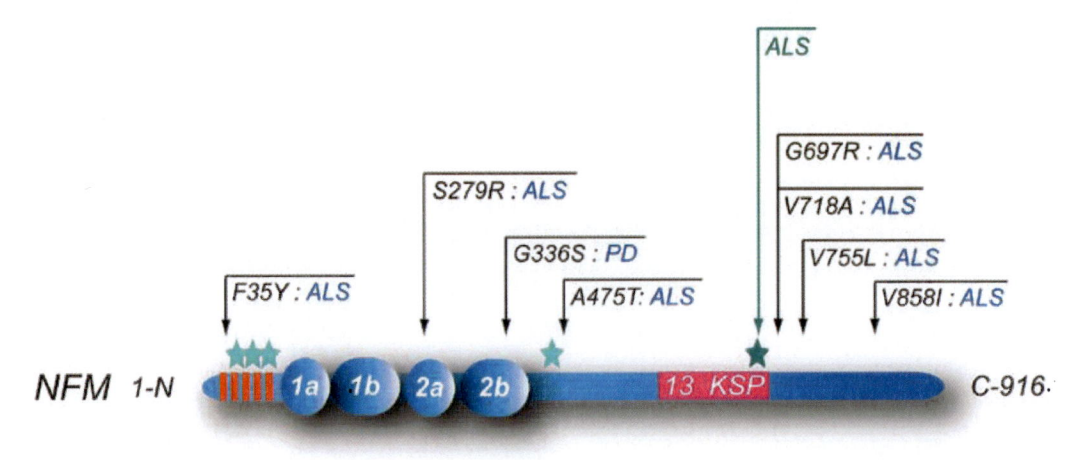

Figure 3. Schematic representation of NFM alterations in various brain diseases. Mutations identified in association with diseases are indicated above the NFM scheme, while the disease effects on NFM are indicated below the NFM scheme. AD: Alzheimer's disease; ALS: amyotrophic lateral sclerosis; CMT1, CMT2: Charcot-Marie-Tooth disease; MMA: methylmalonic aciduria; NH$_4$: hyperammonemia; PA: propionic aciduria; PD: Parkinson's disease.

Figure 4: Schematic representation of NFH alterations in various brain diseases. Mutations identified in association with diseases are indicated above the NFH scheme, while the disease effects on NFH are indicated below the NFH scheme. AD: Alzheimer's disease; ALS: amyotrophic lateral sclerosis; CMT1, CMT2: Charcot-Marie-Tooth disease; i: insertion; MMA: methylmalonic aciduria; PA: propionic aciduria; PD: Parkinson's disease; Δ: deletion.

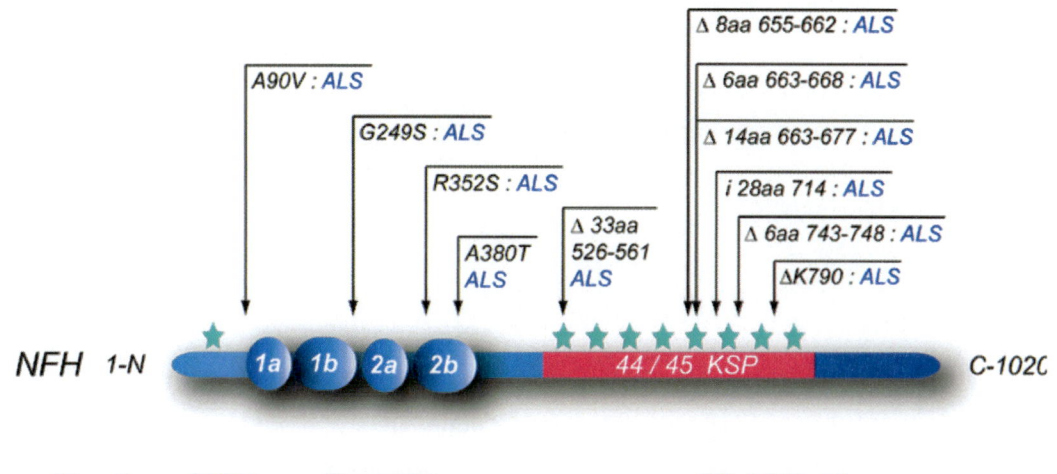

Parkinson Disease (PD)

PD is a progressive neurodegenerative CNS disorder affecting dopaminergic neurons of substantia nigra and leading to decreased dopamine availability. The principal pathological modifications in PD affected neurons are the so-called Lewy bodies, which are inclusions of accumulated proteins in neuronal perikarya and are made of numerous proteins, including NFL, NFM and NFH, α-synuclein, ubiquitin and subunits of the proteasome (Galloway et al., 1992; Trimmer et al., 2004). In particular, abnormally phosphorylated NFs have been identified in PD associated Lewy bodies (Hill et al., 1991; Trojanowski et al., 1993), but the reasons for this alteration of NF phosphorylation have not been precisely identified so far (Figures 2, 3, 4). Familial forms of PD have also been identified, in which the principal mutations found are located in the parkin, α-synuclein and ubiquitin C-terminal hydrolase L1, all three related to cellular ubiquitin proteasomal system (Lim et al., 2003). A significative decrease of NF mRNAs and proteins has also been observed in the PD affected neurons of substantia nigra (Hill et al., 1993; Basso et al., 2004) (Figures 2, 3, 4). Recently, a point mutation in the *NEFM* gene, located in the rod domain 2b of NFM and changing Gly to Ser (Gly336Ser) (Figure 3), has been identified in a patient that developed PD very early, at the age of 16 (Lavedan et al., 2002). Due to the position of this mutation in the very highly conserved region of IFs (rod, α-helical coils) involved in their assembly mechanism, it was speculated that this mutation could alter NFM assembly into NF filaments (Lavedan et al., 2002). As this mutation has been found in only one PD patient which moreover had three unaffected siblings (Lavedan et al., 2002; Han et al., 2005), it is not sure so far that this mutation is really causative of PD. If yes however, the NFM G336S mutation does not seem to interfere with either assembly nor cellular distribution of NFs (Perez-Olle et al., 2004), but could rather alter interactions of NFM with other PD susceptibility proteins (Al-Chalabi and Miller, 2003).

Alzheimer's Disease (AD)

Among neurodegenerative diseases, AD is the leading cause of dementia, with risks over 65 years of age varying from 6-10% for men to 12-19% for women (Seshadri et al., 1997). CNS regions involved in memory and thinking skills are the first affected, followed by neuronal death in other brain regions as disease progresses, which eventually causes the death of the patient. Despite intensive work on AD, its precise cause is still unknown. One of the important secondary features of AD is the neuronal cytoskeleton disruption, due to the inappropriate hyperphosphorylation of cytoskeletal proteins such as tau or NFs (Sternberger et al., 1985; Gong et al., 2000) (Figures 3, 4). In particular, hyperphosphorylated NFH accumulates in neuronal perikaryon and proximal axon (Sternberger et al., 1985), due most probably to an imbalance between kinase and phosphatase activities (Trojanowski et al., 1993; Maccioni et al., 2001; Veeranna et al., 2004). After accumulation in neuronal perikarya, these cytoskeletal proteins aggregate in abnormally modified filaments, and progressively form the neurofibrillary tangles and AD senile plaques, which are the hallmarks of AD. Recently, hyperphosphorylated NFM has also been identified in AD amyloid plaques (Liao et al., 2004). NFL mRNA is also significantly decreased in AD degenerating neurons (McLachlan et al., 1988) (Figure 2).

NFs in Other Neurodegenerative Diseases

The expression and post-translational modifications of NFs have been found altered in a number of other neurodegenerative conditions (summarized in figures 2, 3).

Giant axonal neuropathy (GAN) is a rare autosomal recessive neurodegenerative disorder progressively affecting both peripheral and central nervous system. GAN is due to mutations of the gene encoding gigaxonin, a protein suggested to be associated to IFs (Bomont et al., 2000; Herrmann and Griffin, 2002). GAN, due to the gigaxonin disruption, is thus characterized by the presence of giant axons filled with massive segmental accumulations of disorganized NFs (Asbury et al., 1972; Herguner et al., 2005).

A recent work has shown that leprous nerve atrophy, characterized by a diminution of axonal calibre and paranodal demyelination, might be due to dephosphorylation of NFM and NFH (Save et al., 2004).

NFH have been shown to be dephosphorylated in an experimental model of glaucoma, a neurodegenerative condition affecting the optic nerve in association with high intraoccular pressure (Kashiwagi et al., 2003).

Glutamate excitotoxicity induces a rapid degradation of the neuronal cytoskeleton. It was shown recently that glutamate toxicity, primarily mediated by NMDA receptor, initiates a rapid loss of NFs in the affected axons, while other axonal markers remain intact for a longer period (Chung et al., 2005).

Distal hereditary motor neuronopathies (dHMNs) are a heterogeneous group of disorders in which motor neurons selectively undergo age-dependent degeneration. Mutations in the small heat-shock protein HSPB1 (also called HSP27) are responsible for one form of dHMN. The mutant forms of HSPB1 seem to disrupt NF assembly, to alter axonal transport system, and lead to the accumulation and aggregation, in neuronal perikarya, of cellular components, including NFM (Ackerley et al., 2006).

Huntington's disease (HD) is caused by a polyglutamine repeat expansion in the N-terminal domain of the huntingtin protein. Huntingtin is localized in the cytoplasm where it may interact with cytoskeletal and synaptic proteins. The mechanism of HD pathogenesis remains unknown but recent investigations suggest that the mutant huntingtin found in HD might interact aberrantly with cytoskeletal proteins, including NFs, and thus affect the axonal cytoskeletal integrity (DiProspero et al., 2004).

Neuronal intermediate filament inclusion disease (NIFID) is a recently described novel neurological disease of early onset, presenting considerable variability in clinical phenotypes, including frontotemporal dementia, as well as pyramidal and extrapyramidal signs. The pathological hallmark of NIFID is the presence of abnormal aggregates of α-internexin, NFL, NFM and NFH in the affected neurons (Cairns et al., 2004). α-internexin, a class IV IF protein, has not been identified in any pathological protein aggregates of any other neurodegenerative disease.

NFs in Neurodevelopmental and Metabolic Diseases

Diabetes Neuropathy

Diabetes is associated with a symmetrical distal axonal neuropathy predominantly affecting sensory nerves and neurons of dorsal root ganglia. Diabetic neuropathy is characterized by a reduced conduction velocity, and axonal atrophy. Both in human diabetic patients and in streptozotocin-induced diabetic rats, abnormal aggregations of NFs and other cytoskeletal proteins have been observed in the affected neurons, together with an abnormal increase of NFM and NFH phosphorylation (Figures 3, 4) (Schmidt et al., 1997; Fernyhough et al., 1999). These alterations of NF phosphorylation seem to occur through the activation of the NF kinase c-Jun N-terminal kinase (JNK) (Fernyhough et al., 1999; Middlemas et al., 2006). NFs mRNAs are reduced. The affected neurons present defects of axonal transport mechanisms, a reduction in axon calibre, and a diminished capacity of nerve regeneration, all characteristics relying on the integrity of axonal cytoskeleton. It appears thus that NF abnormalities seem to be a primary cause of diabetic neuropathy, and not only a marker of the pathology (McLean, 1997). A recent work has shown that diabetic neuropathy in an experimental model, the insulin KO mouse, does not alter only peripheral axons, but also

Hyperammonemia during CNS Development

Poorly understood irreversible damages to CNS development occur in neonates and infants with hepatic deficiency or inherited defects of ammonium (NH_4^+) metabolism, manifesting on the long term as mental retardation (Bachmann, 2002; Bachmann, 2003). We have shown, in brain cell 3D primary cultures exposed to NH_4^+ as experimental model of hyperammonemia during CNS development, that NH_4^+ impairs axonal growth (Braissant et al., 1999; Braissant et al., 2002). NFs appear to be affected in this process, as both NFM expression and phosphorylation are decrease by NH_4^+ exposure (Figure 2) (Braissant et al., 2002). The correct expression and phosphorylation of NFM seem to depend on levels of creatine (Braissant et al., 2002), which can be synthesized by brain cells including during development (Braissant et al., 2001; Braissant et al., 2005). Axonal growth, as well as NFM expression and phosphorylation, are protected under NH_4^+ exposure by co-treatment with creatine in a glial cell dependent manner (Braissant et al., 2002). Our results are consistent with clinical findings in hyperammonemic neonates or infants presenting irreversible brain lesions compatible with neuronal fiber loss or defects of neurite outgrowth. The alteration of NF phosphorylation under NH_4^+ exposure might occur through the dysregulation of MAPK, which are NF kinases and present altered levels of phosphorylation and activity in brain cells exposed to NH_4^+ (Schliess et al., 2002; Jayakumar et al., 2006; Cagnon et al., 2006).

Methylmalonic (MMA) and Propionic (PA) Acidemias

Among the most frequent organic acidemias, PA and MMA are due to deficiencies in propionyl-CoA carboxylase and L-methylmalonyl-CoA mutase, respectively, and lead to the increase of free propionic acid in blood and its accumulation in tissues (PA), and to the tissular accumulation of L-methylmalonic acid and secondarily of propionic acid (MMA). The levels of these metabolites in blood and cerebrospinal fluid can rise as high as 5 mM and may be even higher in neuronal cells. PA and MMA lead to chronic neurologic disabilities, seizures and developmental delay. Damages to basal ganglia, a general hypomyelination, cerebral atrophy and white matter edema are frequently encountered. So far, the exact underlying mechanisms of brain damage in PA and MMA remain to be elucidated. However, NFs might be implicated in the neuropathological aspects of MMA and PA (Figures 2, 3 4). Indeed, MMA and PA experimental models have provided evidence that neuronal NFL and NFM expression and phosphorylation are reduced under L-methylmalonic acid and propionic acid exposures (de Mattos-Dutra et al., 1997a; de Mattos-Dutra et al., 1997b; de Mattos-Dutra et al., 1998), while they are increased for NFH (Vivian et al., 2002).

NFs in other Neurodevelopomental and Metabolic Diseases

Phenylketonuria (PKU) is one of the most frequent inborn errors of metabolism, is due to the deficiency of the hepatic enzyme phenylalanine hydroxylase and results in hyperphenylalaninemia. Among other pathological characteristics, untreated PKU leads to mental retardation. Untreated PKU patients show a severe hypomyelination of their CNS. Experimental evidence has been shown that hyperphenylalaninemia delays axonal maturation and myelination during critical period of CNS development, probably through a deficit of NFH as well as myelin basic protein expression (Reynolds et al., 1993).

Progressive encephalopathy syndrome with edema, hypsarrhythmia and optic atrophy (PEHO syndrome) is a form of infantile progressive encephalopathy showing severe hypotonia, convulsions, profound mental retardation, hyperreflexia, optic atrophy and brain atrophy, in particular in cerebellum and brainstem. PEHO seems to occur in the postnatal period, without exclusion of potential prenatal onset. Interestingly, PEHO patients presented an aberrant expression of NFH in the perikarya of their cerebellar Purkinje cells, demonstrating an important disorganization of their cytoskeleton (Haltia and Somer, 1993).

NFs as Markers of Diseases

NFs, as the principal components of the axonal cytoskeleton, are released in the interstitial fluid after axonal injury or degeneration, and diffuse into cerebrospinal fluid (CSF), where they can be quantified to monitor axonal degeneration, as well as disease activity and progression. Increasing studies are published making use of NFs as markers of neuronal injury. A lot of work has been done on the measure of NFL and NFH released in CSF, as markers of axonal degeneration, to help the prediction and monitoring of the neurological decline in people with multiple sclerosis (MS). Different studies have shown that NFL CSF concentration is higher in patients with MS than in controls, making of NFL a promising marker to discriminate MS patients from patients with other neurological diseases. On the other hand, CSF NFH seems interesting for the follow up of the progression of the disease in MS patients, as it is increased during the progressive phase of MS. For more specific informations on the use of NFs as markers of MS, the reader is invited to read two detailed and recent reviews (Petzold, 2005; Teunissen et al., 2005).

As new but non-exhaustive examples, the use of NFs as markers of three other neuropathological conditions will be briefly discussed here: ALS, subarachnoid hemorrhage (SAH), and brain damages as consequence of cardiac arrest.

As discussed in a previous chapter, ALS is the most common form of motor neuron disease, presenting as neuropathological hallmark an abnormal aggregation of NFs in the degenerating motor neurons. A recent work proposes that phosphorylated NFH might be a valuable marker of axonal damage in ALS, discriminate between different categories of ALS, and be used as marker for therapeutic trials (Brettschneider et al., 2006).

Axonal degeneration is thought to be an underestimated complication of SAH, which can continue for days after the primary injury, and extend into the period of delayed cerebral ischemia. A recent study shows that phosphorylated NFH, measured daily in CSF during 14 days after the SAH episode, is significatively increased in SAH patients with bad outcome (measured at 3 months) (Petzold et al., 2005). This work demonstrates the secondary axonal degeneration following SAH, and show that the levels of phosphorylated NFH in CSF are highly predictive of a bad outcome for SAH patients.

The majority of patients surviving resuscitation after an out of hospital cardiac arrest present neurological complications due to global anoxia. Outcome prediction for these patients mainly rely on clinical observations, and on the recent measure of biochemical markers of brain damage in serum, such as brain specific proteins S-100 or NSE (Rosen et al., 2001). A recent study has shown that the levels of NFL in CSF give a reliable measure of brain damage, and are highly predictive of poor outcome for these patients (Rosen et al., 2004).

Conclusion

NFs are essential cytoskeletal proteins of the neuron, which participate in axonal rigidity, tensile strength, stability along time, regulation of calibre, and transport guidance of organelles and particles. NFs alterations have been identified in many different brain pathologies, ranging from neurodegenerative, neurodevelopmental to metabolic diseases. This list of diseases showing abnormalities in NFs will certainly increase in the near future. The identified NFs alterations range from genetic mutations, to abnormal expression, post-translational modifications and aberrant localization or accumulation in neuronal perikaryion. From this diversity of NF dysregulation in so many brain diseases, the future experimental work on NFs may unravel common mechanisms of IF accumulation and aggregation, and hopefully allow the design of better treatments for the patients suffering of these neurodegenerative diseases.

Acknowledgments

Our work is supported by the Swiss National Science Foundation, grants n° 31-63892.00 and 3100A0-100778.

References

Ackerley S, James PA, Kalli A, French S, Davies KE, and Talbot K (2006). A mutation in the small heat-shock protein HSPB1 leading to distal hereditary motor neuronopathy disrupts neurofilament assembly and the axonal transport of specific cellular cargoes. *Hum. Mol. Genet. 15*, 347-354.

Al-Chalabi A, Andersen PM, Nilsson P, Chioza B, Andersson JL, Russ C, Shaw CE, Powell JF, and Leigh PN (1999). Deletions of the heavy neurofilament subunit tail in amyotrophic lateral sclerosis. *Hum. Mol. Genet. 8*, 157-164.

Al-Chalabi A and Miller CC (2003). Neurofilaments and neurological disease. *Bioessays 25*, 346-355.

Andersen PM (2006). Amyotrophic lateral sclerosis associated with mutations in the CuZn superoxide dismutase gene. *Curr. Neurol. Neurosci. Rep. 6*, 37-46.

Asbury AK, Gale MK, Cox SC, Baringer JR, and Berg BO (1972). Giant axonal neuropathy - a unique case with segmental neurofilamentous masses. *Acta Neuropathol. 20*, 237-247.

Baas PW (1997). Microtubules and axonal growth. *Curr. Opin. Cell Biol. 9*, 29-36.

Bachmann C (2002). Mechanisms of hyperammonemia. Clin. Chem. Lab. Med. *40*, 653-662.

Bachmann C (2003). Outcome and survival of 88 patients with urea cycle disorders: a retrospective evaluation. *Eur. J Pediatr. 162*, 410-416.

Basso M, Giraudo S, Corpillo D, Bergamasco B, Lopiano L, and Fasano M (2004). Proteome analysis of human substantia nigra in Parkinson's disease. *Proteomics. 4*, 3943-3952.

Beaulieu JM, Robertson J, and Julien JP (1999). Interactions between peripherin and neurofilaments in cultured cells: disruption of peripherin assembly by the NF-M and NF-H subunits. *Biochem. Cell Biol. 77*, 41-45.

Bergeron C, Beric-Maskarel K, Muntasser S, Weyer L, Somerville MJ, and Percy ME (1994). Neurofilament light and polyadenylated mRNA levels are decreased in amyotrophic lateral sclerosis motor neurons. *J. Neuropathol. Exp. Neurol. 53*, 221-230.

Bignami A, Raju T, and Dahl D (1982). Localization of vimentin, the nonspecific intermediate filament protein, in embryonal glia and in early differentiating neurons. In vivo and in vitro immunofluorescence study of the rat embryo with vimentin and neurofilament antisera. *Dev. Biol. 91*, 286-295.

Bomont P, Cavalier L, Blondeau F, Ben HC, Belal S, Tazir M, Demir E, Topaloglu H, Korinthenberg R, Tuysuz B, Landrieu P, Hentati F, and Koenig M (2000). The gene encoding gigaxonin, a new member of the cytoskeletal BTB/kelch repeat family, is mutated in giant axonal neuropathy. *Nat. Genet. 26*, 370-374.

Braissant O, Henry H, Loup M, Eilers B, and Bachmann C (2001). Endogenous synthesis and transport of creatine in the rat brain: an in situ hybridization study. *Brain Res. Mol. Brain Res. 86*, 193-201.

Braissant O, Henry H, Villard AM, Speer O, Wallimann T, and Bachmann C (2005). *Creatine synthesis and transport during rat embryogenesis: spatiotemporal expression of AGAT, GAMT and CT1.* BMC. Dev. Biol. *5*, 9.

Braissant O, Henry H, Villard AM, Zurich MG, Loup M, Eilers B, Parlascino G, Matter E, Boulat O, Honegger P, and Bachmann C (2002). Ammonium-induced impairment of axonal growth is prevented through glial creatine. *J. Neurosci. 22*, 9810-9820.

Braissant O, Honegger P, Loup M, Iwase K, Takiguchi M, and Bachmann C (1999). Hyperammonemia: regulation of argininosuccinate synthetase and argininosuccinate lyase genes in aggregating cell cultures of fetal rat brain. *Neurosci. Lett. 266*, 89-92.

Brettschneider J, Petzold A, Süssmuth SD, Ludolph AC, and Tumani H (2006). Axonal damage markers in cerebrospinal fluid are increased in ALS. *Neurology 66*, 852-856.

Cagnon L, Honegger P, Bachmann C, and Braissant O (2006). Signal transduction mechanisms implicated in ammonium neurotoxicity and protection through glial creatine. FENS Forum Abstracts 2006. *A228-4*.

Cairns NJ, Zhukareva V, Uryu K, Zhang B, Bigio E, Mackenzie IR, Gearing M, Duyckaerts C, Yokoo H, Nakazato Y, Jaros E, Perry RH, Lee VM, and Trojanowski JQ (2004). alpha-internexin is present in the pathological inclusions of neuronal intermediate filament inclusion disease. *Am. J. Pathol. 164*, 2153-2161.

Carden MJ, Trojanowski JQ, Schlaepfer WW, and Lee VM (1987). Two-stage expression of neurofilament polypeptides during rat neurogenesis with early establishment of adult phosphorylation patterns. *J. Neurosci. 7*, 3489-3504.

Chen J, Nakata T, Zhang Z, and Hirokawa N (2000). The C-terminal tail domain of neurofilament protein-H (NF-H) forms the crossbridges and regulates neurofilament bundle formation. *J. Cell Sci. 113*, 3861-3869.

Chin SS, Macioce P, and Liem RK (1991). Effects of truncated neurofilament proteins on the endogenous intermediate filaments in transfected fibroblasts. *J. Cell Sci. 99*, 335-350.

Ching GY and Liem RK (1999). Analysis of the roles of the head domains of type IV rat neuronal intermediate filament proteins in filament assembly using domain-swapped chimeric proteins. *J. Cell Sci. 112*, 2233-2240.

Ching GY and Liem RK (1993). Assembly of type IV neuronal intermediate filaments in nonneuronal cells in the absence of preexisting cytoplasmic intermediate filaments. *J. Cell Biol. 122*, 1323-1335.

Choi BO, Lee MS, Shin SH, Hwang JH, Choi KG, Kim WK, Sunwoo IN, Kim NK, and Chung KW (2004). Mutational analysis of PMP22, MPZ, GJB1, EGR2 and NEFL in Korean Charcot-Marie-Tooth neuropathy patients. *Hum. Mutat. 24*, 185-186.

Chung RS, McCormack GH, King AE, West AK, and Vickers JC (2005). Glutamate induces rapid loss of axonal neurofilament proteins from cortical neurons in vitro. *Exp. Neurol. 193*, 481-488.

Cleverley KE, Betts JC, Blackstock WP, Gallo JM, and Anderton BH (1998). Identification of novel in vitro PKA phosphorylation sites on the low and middle molecular mass neurofilament subunits by mass spectrometry. *Biochemistry 37*, 3917-3930.

Cochard P and Paulin D (1984). Initial expression of neurofilaments and vimentin in the central and peripheral nervous system of the mouse embryo in vivo. *J. Neurosci. 4*, 2080-2094.

Cohlberg JA, Hajarian H, Tran T, Alipourjeddi P, and Noveen A (1995). Neurofilament protein heterotetramers as assembly intermediates. *J. Biol. Chem. 270*, 9334-9339.

De Mattos-Dutra A, Sampaio de Freitas M, Lisboa CS, Pessoa-Pureur R, and Wajner M (1998). Effects of acute and chronic administration of methylmalonic and propionic acids on the in vitro incorporation of 32P into cytoskeletal proteins from cerebral cortex of young rats. *Neurochem. Int. 33*, 75-82.

De Mattos-Dutra A, Sampaio de Freitas M, Schröder N, Zilles AC, Wajner M, and Pessoa-Pureur R (1997a). Methylmalonic acid reduces the in vitro phosphorylation of cytoskeletal proteins in the cerebral cortex of rats. *Brain Res. 763*, 221-231.

De Mattos-Dutra A, Sampaio de Freitas M, Schröder N, Fogaca Lisboa CS, Pessoa-Pureur R, and Wajner M (1997b). In vitro phosphorylation of cytoskeletal proteins in the rat cerebral cortex is decreased by propionic acid. *Exp. Neurol. 147*, 238-247.

De Waegh SM, Lee VM, and Brady ST (1992). Local modulation of neurofilament phosphorylation, axonal caliber, and slow axonal transport by myelinating Schwann cells. *Cell 68*, 451-463.

De Jonghe P, Mersivanova I, Nelis E, Del Favero J, Martin JJ, Van Broeckhoven C, Evgrafov O, and Timmerman V (2001). Further evidence that neurofilament light chain gene mutations can cause Charcot-Marie-Tooth disease type 2E. *Ann. Neurol. 49*, 245-249.

DiProspero NA, Chen EY, Charles V, Plomann M, Kordower JH, and Tagle DA (2004). Early changes in Huntington's disease patient brains involve alterations in cytoskeletal and synaptic elements. *J. Neurocytol. 33*, 517-533.

Dong DL, Xu ZS, Chevrier MR, Cotter RJ, Cleveland DW, and Hart GW (1993). Glycosylation of mammalian neurofilaments. Localization of multiple O-linked N-acetylglucosamine moieties on neurofilament polypeptides L and M. *J. Biol. Chem. 268*, 16679-16687.

Dong DL, Xu ZS, Hart GW, and Cleveland DW (1996). Cytoplasmic O-GlcNAc modification of the head domain and the KSP repeat motif of the neurofilament protein neurofilament-H. *J. Biol. Chem. 271*, 20845-20852.

Elder GA, Friedrich VL Jr, Bosco P, Kang C, Gourov A, Tu PH, Lee VM, and Lazzarini RA (1998a). Absence of the mid-sized neurofilament subunit decreases axonal calibers, levels of light neurofilament (NF-L), and neurofilament content. *J. Cell Biol. 141*, 727-739.

Elder GA, Friedrich VL Jr, Kang C, Bosco P, Gourov A, Tu PH, Zhang B, Lee VM, and Lazzarini RA (1998b). Requirement of heavy neurofilament subunit in the development of axons with large calibers. *J. Cell Biol. 143*, 195-205.

Elder GA, Friedrich VL Jr, Margita A, and Lazzarini RA (1999a). Age-related atrophy of motor axons in mice deficient in the mid-sized neurofilament subunit. *J. Cell Biol. 146*, 181-192.

Elder GA, Friedrich VL Jr, Pereira D, Tu PH, Zhang B, Lee VM, and Lazzarini RA (1999b). Mice with disrupted midsized and heavy neurofilament genes lack axonal neurofilaments but have unaltered numbers of axonal microtubules. *J. Neurosci. Res. 57*, 23-32.

Fernyhough P, Gallagher A, Averill SA, Priestley JV, Hounsom L, Patel J, and Tomlinson DR (1999). Aberrant neurofilament phosphorylation in sensory neurons of rats with diabetic neuropathy. *Diabetes 48*, 881-889.

Figlewicz DA, Krizus A, Martinoli MG, Meininger V, Dib M, Rouleau GA, and Julien JP (1994). Variants of the heavy neurofilament subunit are associated with the development of amyotrophic lateral sclerosis. *Hum. Mol. Genet. 3*, 1757-1761.

Figlewicz DA, Rouleau GA, Krizus A, and Julien JP (1993). Polymorphism in the multi-phosphorylation domain of the human neurofilament heavy-subunit-encoding gene. *Gene 132*, 297-300.

Fliegner KH, Kaplan MP, Wood TL, Pintar JE, and Liem RK (1994). Expression of the gene for the neuronal intermediate filament protein alpha-internexin coincides with the onset of neuronal differentiation in the developing rat nervous system. *J. Comp Neurol. 342*, 161-173.

Fuchs E and Cleveland DW (1998). A structural scaffolding of intermediate filaments in health and disease. *Science 279*, 514-519.

Fuchs E and Weber K (1994). Intermediate filaments: structure, dynamics, function, and disease. *Annu. Rev. Biochem. 63*, 345-382.

Galloway PG, Mulvihill P, and Perry G (1992). Filaments of Lewy bodies contain insoluble cytoskeletal elements. *Am. J. Pathol. 140*, 809-822.

Garcia ML, Singleton AB, Hernandez D, Ward CM, Evey C, Sapp PA, Hardy J, Brown RH Jr, and Cleveland DW (2006). Mutations in neurofilament genes are not a significant primary cause of non-SOD1-mediated amyotrophic lateral sclerosis. *Neurobiol. Dis. 21*, 102-109.

Ge WW, Wen W, Strong W, Leystra-Lantz C, and Strong MJ (2005). Mutant copper-zinc superoxide dismutase binds to and destabilizes human low molecular weight neurofilament mRNA. *J. Biol. Chem. 280*, 118-124.

Georgiou DM, Zidar J, Korosec M, Middleton LT, Kyriakides T, and Christodoulou K (2002). A novel NF-L mutation Pro22Ser is associated with CMT2 in a large Slovenian family. *Neurogenetics. 4*, 93-96.

Giasson BI and Mushynski WE (1996). Aberrant stress-induced phosphorylation of perikaryal neurofilaments. *J. Biol. Chem. 271*, 30404-30409.

Gibb BJ, Brion JP, Brownlees J, Anderton BH, and Miller CC (1998). Neuropathological abnormalities in transgenic mice harbouring a phosphorylation mutant neurofilament transgene. *J. Neurochem. 70*, 492-500.

Gibb BJ, Robertson J, and Miller CC (1996). Assembly properties of neurofilament light chain Ser55 mutants in transfected mammalian cells. *J. Neurochem. 66*, 1306-1311.

Gill SR, Wong PC, Monteiro MJ, and Cleveland DW (1990). Assembly properties of dominant and recessive mutations in the small mouse neurofilament (NF-L) subunit. *J. Cell Biol. 111*, 2005-2019.

Goldstein ME, Sternberger NH, and Sternberger LA (1987). Phosphorylation protects neurofilaments against proteolysis. *J. Neuroimmunol. 14*, 149-160.

Gong CX, Lidsky T, Wegiel J, Zuck L, Grundke-Iqbal I, and Iqbal K (2000). Phosphorylation of microtubule-associated protein tau is regulated by protein phosphatase 2A in mammalian brain. Implications for neurofibrillary degeneration in Alzheimer's disease. *J. Biol. Chem. 275*, 5535-5544.

Gong CX, Wang JZ, Iqbal K, and Grundke-Iqbal I (2003). Inhibition of protein phosphatase 2A induces phosphorylation and accumulation of neurofilaments in metabolically active rat brain slices. *Neurosci. Lett. 340*, 107-110.

Grant P and Pant HC (2000). Neurofilament protein synthesis and phosphorylation. *J. Neurocytol. 29*, 843-872.

Guan RJ, Khatra BS, and Cohlberg JA (1991). Phosphorylation of bovine neurofilament proteins by protein kinase FA (glycogen synthase kinase 3). *J. Biol. Chem. 266*, 8262-8267.

Haltia M and Somer M (1993). Infantile cerebello-optic atrophy. Neuropathology of the progressive encephalopathy syndrome with edema, hypsarrhythmia and optic atrophy (the PEHO syndrome). *Acta Neuropathol. 85*, 241-247.

Han F, Bulman DE, Panisset M, and Grimes DA (2005). Neurofilament M gene in a French-Canadian population with Parkinson's disease. *Can. J. Neurol. Sci. 32*, 68-70.

Heins S and Aebi U (1994). Making heads and tails of intermediate filament assembly, dynamics and networks. *Curr. Opin. Cell Biol. 6*, 25-33.

Heins S, Wong PC, Müller S, Goldie K, Cleveland DW, and Aebi U (1993). The rod domain of NF-L determines neurofilament architecture, whereas the end domains specify filament assembly and network formation. *J. Cell Biol. 123*, 1517-1533.

Herguner MO, Zorludemir S, and Altunbasak S (2005). Giant axonal neuropathy in two siblings: clinical histopathological findings. *Clin. Neuropathol. 24*, 48-50.

Herrmann DN and Griffin JW (2002). Intermediate filaments: a common thread in neuromuscular disorders. *Neurology 58*, 1141-1143.

Herrmann H and Aebi U (2000). Intermediate filaments and their associates: multi-talented structural elements specifying cytoarchitecture and cytodynamics. *Curr. Opin. Cell Biol. 12*, 79-90.

Hill WD, Arai M, Cohen JA, and Trojanowski JQ (1993). Neurofilament mRNA is reduced in Parkinson's disease substantia nigra pars compacta neurons. *J. Comp Neurol. 329*, 328-336.

Hill WD, Lee VM, Hurtig HI, Murray JM, and Trojanowski JQ (1991). Epitopes located in spatially separate domains of each neurofilament subunit are present in Parkinson's disease Lewy bodies. *J. Comp Neurol. 309*, 150-160.

Hirokawa N (1997). The mechanisms of fast and slow transport in neurons: identification and characterization of the new kinesin superfamily motors. *Curr. Opin. Neurobiol. 7*, 605-614.

Hisanaga S, Kusubata M, Okumura E, and Kishimoto T (1991). Phosphorylation of neurofilament H subunit at the tail domain by CDC2 kinase dissociates the association to microtubules. *J. Biol. Chem. 266*, 21798-21803.

Hisanaga S, Matsuoka Y, Nishizawa K, Saito T, Inagaki M, and Hirokawa N (1994). Phosphorylation of native and reassembled neurofilaments composed of NF-L, NF-M, and NF-H by the catalytic subunit of cAMP-dependent protein kinase. *Mol. Biol. Cell 5*, 161-172.

Hsieh ST, Kidd GJ, Crawford TO, Xu Z, Lin WM, Trapp BD, Cleveland DW, and Griffin JW (1994). Regional modulation of neurofilament organization by myelination in normal axons. *J. Neurosci. 14*, 6392-6401.

Jacomy H, Zhu Q, Couillard-Després S, Beaulieu JM, and Julien JP (1999). Disruption of type IV intermediate filament network in mice lacking the neurofilament medium and heavy subunits. *J. Neurochem. 73*, 972-984.

Jayakumar AR, Panickar KS, Murthy C, and Norenberg MD (2006). Oxidative stress and mitogen-activated protein kinase phosphorylation mediate ammonia-induced cell swelling and glutamate uptake inhibition in cultured astrocytes. *J. Neurosci. 26*, 4774-4784.

Jordanova A, De Jonghe P, Boerkoel CF, Takashima H, De Vriendt E, Ceuterick C, Martin JJ, Butler IJ, Mancias P, Papasozomenos SC, Terespolsky D, Potocki L, Brown CW, Shy M, Rita DA, Tournev I, Kremensky I, Lupski JR, and Timmerman V (2003). Mutations in the neurofilament light chain gene (NEFL) cause early onset severe Charcot-Marie-Tooth disease. *Brain 126*, 590-597.

Julien JP and Mushynski WE (1982). Multiple phosphorylation sites in mammalian neurofilament polypeptides. *J. Biol. Chem. 257*, 10467-10470.

Kaplan MP, Chin SS, Fliegner KH, and Liem RK (1990). Alpha-internexin, a novel neuronal intermediate filament protein, precedes the low molecular weight neurofilament protein (NF-L) in the developing rat brain. *J. Neurosci. 10*, 2735-2748.

Kashiwagi K, Ou B, Nakamura S, Tanaka Y, Suzuki M, and Tsukahara S (2003). Increase in dephosphorylation of the heavy neurofilament subunit in the monkey chronic glaucoma model. Invest Ophthalmol. *Vis. Sci. 44*, 154-159.

Kim OJ, Ariano MA, Lazzarini RA, Levine MS, and Sibley DR (2002). Neurofilament-M interacts with the D1 dopamine receptor to regulate cell surface expression and desensitization. *J. Neurosci. 22*, 5920-5930.

Kong J, Tung VW, Aghajanian J, and Xu Z (1998). Antagonistic roles of neurofilament subunits NF-H and NF-M against NF-L in shaping dendritic arborization in spinal motor neurons. *J. Cell Biol. 140*, 1167-1176.

Larivière RC and Julien JP (2004). Functions of intermediate filaments in neuronal development and disease. *J. Neurobiol. 58*, 131-148.

Lavedan C, Buchholtz S, Nussbaum RL, Albin RL, and Polymeropoulos MH (2002). A mutation in the human neurofilament M gene in Parkinson's disease that suggests a role for the cytoskeleton in neuronal degeneration. *Neurosci. Lett. 322*, 57-61.

Lee MK and Cleveland DW (1996). Neuronal intermediate filaments. *Annu. Rev. Neurosci. 19*, 187-217.

Lee MK, Xu Z, Wong PC, and Cleveland DW (1993). Neurofilaments are obligate heteropolymers in vivo. *J. Cell Biol. 122*, 1337-1350.

Lendahl U, Zimmerman LB, and McKay RD (1990). CNS stem cells express a new class of intermediate filament protein. *Cell 60*, 585-595.

Lewis SE and Nixon RA (1988). Multiple phosphorylated variants of the high molecular mass subunit of neurofilaments in axons of retinal cell neurons: characterization and evidence for their differential association with stationary and moving neurofilaments. *J. Cell Biol. 107*, 2689-2701.

Li BS, Daniels MP, and Pant HC (2001). Integrins stimulate phosphorylation of neurofilament NF-M subunit KSP repeats through activation of extracellular regulated-kinases (Erk1/Erk2) in cultured motoneurons and transfected NIH 3T3 cells. *J. Neurochem. 76*, 703-710.

Liao L, Cheng D, Wang J, Duong DM, Losik TG, Gearing M, Rees HD, Lah JJ, Levey AI, and Peng J (2004). Proteomic characterization of postmortem amyloid plaques isolated by laser capture microdissection. *J. Biol. Chem. 279*, 37061-37068.

Lim KL, Dawson VL, and Dawson TM (2003). The cast of molecular characters in Parkinson's disease: felons, conspirators, and suspects. Ann. N. Y. Acad. Sci. *991*, 80-92.

Liu Q, Xie F, Siedlak SL, Nunomura A, Honda K, Moreira PI, Zhua X, Smith MA, and Perry G (2004). Neurofilament proteins in neurodegenerative diseases. *Cell Mol. Life Sci. 61*, 3057-3075.

Lüdemann N, Clement A, Hans VH, Leschik J, Behl C, and Brandt R (2005). O-glycosylation of the tail domain of neurofilament protein M in human neurons and in spinal cord tissue of a rat model of amyotrophic lateral sclerosis (ALS). *J. Biol. Chem. 280*, 31648-31658.

Maccioni RB, Otth C, Concha II, and Munoz JP (2001). The protein kinase Cdk5. Structural aspects, roles in neurogenesis and involvement in Alzheimer's pathology. *Eur. J. Biochem. 268*, 1518-1527.

Manetto V, Sternberger NH, Perry G, Sternberger LA, and Gambetti P (1988). Phosphorylation of neurofilaments is altered in amyotrophic lateral sclerosis. J. Neuropathol. *Exp. Neurol. 47*, 642-653.

Maxwell WL, Povlishock JT, and Graham DL (1997). A mechanistic analysis of nondisruptive axonal injury: a review. *J. Neurotrauma 14*, 419-440.

McLachlan DR, Lukiw WJ, Wong L, Bergeron C, and Bech-Hansen NT (1988). Selective messenger RNA reduction in Alzheimer's disease. Brain Res. *427*, 255-261.

McLean WG (1997). The role of axonal cytoskeleton in diabetic neuropathy. *Neurochem. Res. 22*, 951-956.

Mersiyanova IV, Perepelov AV, Polyakov AV, Sitnikov VF, Dadali EL, Oparin RB, Petrin AN, and Evgrafov OV (2000). A new variant of Charcot-Marie-Tooth disease type 2 is probably the result of a mutation in the neurofilament-light gene. *Am. J. Hum. Genet. 67*, 37-46.

Middlemas AB, Agthong S, and Tomlinson DR (2006). Phosphorylation of c-Jun N-terminal kinase (JNK) in sensory neurones of diabetic rats, with possible effects on nerve conduction and neuropathic pain: prevention with an aldose reductase inhibitor. *Diabetologia 49*, 580-587.

Mukai H, Toshimori M, Shibata H, Kitagawa M, Shimakawa M, Miyahara M, Sunakawa H, and Ono Y (1996). PKN associates and phosphorylates the head-rod domain of neurofilament protein. *J. Biol. Chem. 271*, 9816-9822.

Munoz DG, Greene C, Perl DP, and Selkoe DJ (1988). Accumulation of phosphorylated neurofilaments in anterior horn motoneurons of amyotrophic lateral sclerosis patients. J. Neuropathol. *Exp. Neurol. 47*, 9-18.

Nakamura Y, Hashimoto R, Kashiwagi Y, Aimoto S, Fukusho E, Matsumoto N, Kudo T, and Takeda M (2000). Major phosphorylation site (Ser55) of neurofilament L by cyclic AMP-dependent protein kinase in rat primary neuronal culture. *J. Neurochem. 74*, 949-959.

Nakamura Y, Hashimoto R, Kashiwagi Y, Wada Y, Sakoda S, Miyamae Y, Kudo T, and Takeda M (1999). Casein kinase II is responsible for phosphorylation of NF-L at Ser-473. *FEBS Lett. 455*, 83-86.

Nguyen MD, Larivière RC, and Julien JP (2001). Deregulation of Cdk5 in a mouse model of ALS: toxicity alleviated by perikaryal neurofilament inclusions. *Neuron 30*, 135-147.

Nixon RA (1998). The slow axonal transport debate. *Trends Cell Biol. 8*, 100.

Nixon RA (1993). The regulation of neurofilament protein dynamics by phosphorylation: clues to neurofibrillary pathobiology. *Brain Pathol. 3*, 29-38.

Nixon RA, Paskevich PA, Sihag RK, and Thayer CY (1994). Phosphorylation on carboxyl terminus domains of neurofilament proteins in retinal ganglion cell neurons in vivo: influences on regional neurofilament accumulation, interneurofilament spacing, and axon caliber. *J. Cell Biol. 126*, 1031-1046.

Nixon RA and Shea TB (1992). Dynamics of neuronal intermediate filaments: a developmental perspective. Cell Motil. *Cytoskeleton 22*, 81-91.

Okabe S, Miyasaka H, and Hirokawa N (1993). Dynamics of the neuronal intermediate filaments. *J. Cell Biol. 121*, 375-386.

Pant HC (1988). Dephosphorylation of neurofilament proteins enhances their susceptibility to degradation by calpain. *Biochem. J. 256*, 665-668.

Pant HC and Veeranna (1995). Neurofilament phosphorylation. *Biochem. Cell Biol. 73*, 575-592.

Parry DA and Steinert PM (1999b). Intermediate filaments: molecular architecture, assembly, dynamics and polymorphism. *Q. Rev. Biophys. 32*, 99-187.

Parry DA and Steinert PM (1999a). Intermediate filaments: molecular architecture, assembly, dynamics and polymorphism. *Q. Rev. Biophys. 32*, 99-187.

Perez-Olle R, Lopez-Toledano MA, and Liem RK (2004). The G336S variant in the human neurofilament-M gene does not affect its assembly or distribution: importance of the functional analysis of neurofilament variants. *J. Neuropathol. Exp. Neurol. 63*, 759-774.

Petzold A (2005). Neurofilament phosphoforms: surrogate markers for axonal injury, degeneration and loss. *J. Neurol. Sci. 233*, 183-198.

Petzold A, Rejdak K, Belli A, Sen J, Keir G, Kitchen N, Smith M, and Thompson EJ (2005). Axonal pathology in subarachnoid and intracerebral hemorrhage. *J. Neurotrauma 22*, 407-414.

Portier MM, de Néchaud B, and Gros F (1983). Peripherin, a new member of the intermediate filament protein family. *Dev. Neurosci. 6*, 335-344.

Rao MV, Campbell J, Yuan A, Kumar A, Gotow T, Uchiyama Y, and Nixon RA (2003). The neurofilament middle molecular mass subunit carboxyl-terminal tail domains is essential for the radial growth and cytoskeletal architecture of axons but not for regulating neurofilament transport rate. *J. Cell Biol. 163*, 1021-1031.

Rao MV, Engle LJ, Mohan PS, Yuan A, Qiu D, Cataldo A, Hassinger L, Jacobsen S, Lee VM, Andreadis A, Julien JP, Bridgman PC, and Nixon RA (2002a). Myosin Va binding to neurofilaments is essential for correct myosin Va distribution and transport and neurofilament density. *J. Cell Biol. 159*, 279-290.

Rao MV, Garcia ML, Miyazaki Y, Gotow T, Yuan A, Mattina S, Ward CM, Calcutt NA, Uchiyama Y, Nixon RA, and Cleveland DW (2002b). Gene replacement in mice reveals that the heavily phosphorylated tail of neurofilament heavy subunit does not affect axonal caliber or the transit of cargoes in slow axonal transport. *J. Cell Biol. 158*, 681-693.

Rao MV, Houseweart MK, Williamson TL, Crawford TO, Folmer J, and Cleveland DW (1998). Neurofilament-dependent radial growth of motor axons and axonal organization of neurofilaments does not require the neurofilament heavy subunit (NF-H) or its phosphorylation. *J. Cell Biol. 143*, 171-181.

Reilly MM (2000). Classification of the hereditary motor and sensory neuropathies 7. *Curr. Opin. Neurol. 13*, 561-564.

Reynolds R, Burri R, and Herschkowitz N (1993). Retarded development of neurons and oligodendroglia in rat forebrain produced by hyperphenylalaninemia results in permanent deficits in myelin despite long recovery periods. *Exp. Neurol. 124*, 357-367.

Rosen H, Karlsson JE, and Rosengren L (2004). CSF levels of neurofilament is a valuable predictor of long-term outcome after cardiac arrest. *J. Neurol. Sci. 221*, 19-24.

Rosen H, Sunnerhagen KS, Herlitz J, Blomstrand C, and Rosengren L (2001). Serum levels of the brain-derived proteins S-100 and NSE predict long-term outcome after cardiac arrest. *Resuscitation 49*, 183-191.

Rouleau GA, Clark AW, Rooke K, Pramatarova A, Krizus A, Suchowersky O, Julien JP, and Figlewicz D (1996). SOD1 mutation is associated with accumulation of neurofilaments in amyotrophic lateral sclerosis. *Ann. Neurol. 39*, 128-131.

Roy S, Coffee P, Smith G, Liem RK, Brady ST, and Black MM (2000). Neurofilaments are transported rapidly but intermittently in axons: implications for slow axonal transport. *J. Neurosci. 20*, 6849-6861.

Sanchez I, Hassinger L, Paskevich PA, Shine HD, and Nixon RA (1996). Oligodendroglia regulate the regional expansion of axon caliber and local accumulation of neurofilaments during development independently of myelin formation. *J. Neurosci. 16*, 5095-5105.

Sanchez I, Hassinger L, Sihag RK, Cleveland DW, Mohan P, and Nixon RA (2000). Local control of neurofilament accumulation during radial growth of myelinating axons in vivo. Selective role of site-specific phosphorylation. *J. Cell Biol. 151*, 1013-1024.

Save MP, Shetty VP, Shetty KT, and Antia NH (2004). Alterations in neurofilament protein(s) in human leprous nerves: morphology, immunohistochemistry and Western immunoblot correlative study. Neuropathol. *Appl. Neurobiol. 30*, 635-650.

Schechter R, Beju D, and Miller KE (2005). The effect of insulin deficiency on tau and neurofilament in the insulin knockout mouse. *Biochem. Biophys. Res. Commun. 334*, 979-986.

Schliess F, Gorg B, Fischer R, Desjardins P, Bidmon HJ, Herrmann A, Butterworth RF, Zilles K, and Häussinger D (2002). Ammonia induces MK-801-sensitive nitration and phosphorylation of protein tyrosine residues in rat astrocytes. *FASEB J. 16*, 739-741.

Schmidt RE, Beaudet LN, Plurad SB, and Dorsey DA (1997). Axonal cytoskeletal pathology in aged and diabetic human sympathetic autonomic ganglia. *Brain Res. 769*, 375-383.

Seshadri S, Wolf PA, Beiser A, Au R, McNulty K, White R, and D'Agostino RB (1997). Lifetime risk of dementia and Alzheimer's disease. The impact of mortality on risk estimates in the Framingham Study. *Neurology 49*, 1498-1504.

Shah JV, Flanagan LA, Janmey PA, and Leterrier JF (2000). Bidirectional translocation of neurofilaments along microtubules mediated in part by dynein/dynactin. *Mol. Biol. Cell 11*, 3495-3508.

Shaw G and Weber K (1982). Differential expression of neurofilament triplet proteins in brain development. *Nature 298*, 277-279.

Shea TB, Jung C, and Pant HC (2003). Does neurofilament phosphorylation regulate axonal transport? *Trends Neurosci. 26*, 397-400.

Sihag RK and Nixon RA (1989). In vivo phosphorylation of distinct domains of the 70-kilodalton neurofilament subunit involves different protein kinases. *J. Biol. Chem. 264*, 457-464.

Sihag RK and Nixon RA (1991). Identification of Ser-55 as a major protein kinase A phosphorylation site on the 70-kDa subunit of neurofilaments. Early turnover during axonal transport. *J. Biol. Chem. 266*, 18861-18867.

Skre H (1974). Genetic and clinical aspects of Charcot-Marie-Tooth's disease 1. *Clin. Genet.* 6, 98-118.

Sternberger LA and Sternberger NH (1983). Monoclonal antibodies distinguish phosphorylated and nonphosphorylated forms of neurofilaments in situ. *Proc. Natl. Acad. Sci. U. S. A 80*, 6126-6130.

Sternberger NH, Sternberger LA, and Ulrich J (1985). Aberrant neurofilament phosphorylation in Alzheimer disease. *Proc. Natl. Acad. Sci. U. S. A 82*, 4274-4276.

Sun D, Leung CL, and Liem RK (1996). Phosphorylation of the high molecular weight neurofilament protein (NF-H) by Cdk5 and p35. *J. Biol. Chem. 271*, 14245-14251.

Teunissen CE, Dijkstra C, and Polman C (2005). Biological markers in CSF and blood for axonal degeneration in multiple sclerosis. *Lancet Neurol. 4*, 32-41.

Tomkins J, Usher P, Slade JY, Ince PG, Curtis A, Bushby K, and Shaw PJ (1998). Novel insertion in the KSP region of the neurofilament heavy gene in amyotrophic lateral sclerosis (ALS). *Neuroreport 9*, 3967-3970.

Trimmer PA, Borland MK, Keeney PM, Bennett JP Jr, and Parker WD Jr (2004). Parkinson's disease transgenic mitochondrial cybrids generate Lewy inclusion bodies. *J. Neurochem. 88*, 800-812.

Trojanowski JQ, Schmidt ML, Shin RW, Bramblett GT, Rao D, and Lee VM (1993). Altered tau and neurofilament proteins in neuro-degenerative diseases: diagnostic implications for Alzheimer's disease and Lewy body dementias. *Brain Pathol. 3*, 45-54.

Veeranna, Kaji T, Boland B, Odrljin T, Mohan P, Basavarajappa BS, Peterhoff C, Cataldo A, Rudnicki A, Amin N, Li BS, Pant HC, Hungund BL, Arancio O, and Nixon RA (2004). Calpain mediates calcium-induced activation of the erk1,2 MAPK pathway and cytoskeletal phosphorylation in neurons: relevance to Alzheimer's disease. *Am. J. Pathol. 165*, 795-805.

Vivian L, Pessutto FD, de Almeida LM, Loureiro SO, Pelaez PL, Funchal C, Wajner M, and Pessoa-Pureur R (2002). Effect of propionic and methylmalonic acids on the high molecular weight neurofilament subunit (NF-H) in rat cerebral cortex. *Neurochem. Res. 27*, 1691-1697.

Wang L, Ho CL, Sun D, Liem RK, and Brown A (2000). Rapid movement of axonal neurofilaments interrupted by prolonged pauses. *Nat. Cell Biol. 2*, 137-141.

Wataya T, Nunomura A, Smith MA, Siedlak SL, Harris PL, Shimohama S, Szweda LI, Kaminski MA, Avila J, Price DL, Cleveland DW, Sayre LM, and Perry G (2002). High molecular weight neurofilament proteins are physiological substrates of adduction by the lipid peroxidation product hydroxynonenal. *J. Biol. Chem. 277*, 4644-4648.

Watson DF, Nachtman FN, Kuncl RW, and Griffin JW (1994). Altered neurofilament phosphorylation and beta tubulin isotypes in Charcot-Marie-Tooth disease type 1. *Neurology 44*, 2383-2387.

Willard M and Simon C (1983). Modulations of neurofilament axonal transport during the development of rabbit retinal ganglion cells. *Cell 35*, 551-559.

Williamson TL and Cleveland DW (1999). Slowing of axonal transport is a very early event in the toxicity of ALS-linked SOD1 mutants to motor neurons. *Nat. Neurosci. 2*, 50-56.

Wong NK, He BP, and Strong MJ (2000). Characterization of neuronal intermediate filament protein expression in cervical spinal motor neurons in sporadic amyotrophic lateral sclerosis (ALS). *J. Neuropathol. Exp. Neurol. 59*, 972-982.

Wong PC and Cleveland DW (1990). Characterization of dominant and recessive assembly-defective mutations in mouse neurofilament NF-M. *J. Cell Biol. 111*, 1987-2003.

Xu Z, Dong DL, and Cleveland DW (1994). Neuronal intermediate filaments: new progress on an old subject. *Curr. Opin. Neurobiol. 4*, 655-661.

Xu Z and Tung VW (2000). Overexpression of neurofilament subunit M accelerates axonal transport of neurofilaments. *Brain Res. 866*, 326-332.

Yabe JT, Chylinski T, Wang FS, Pimenta A, Kattar SD, Linsley MD, Chan WK, and Shea TB (2001). Neurofilaments consist of distinct populations that can be distinguished by C-terminal phosphorylation, bundling, and axonal transport rate in growing axonal neurites. *J. Neurosci. 21*, 2195-2205.

Yabe JT, Jung C, Chan WK, and Shea TB (2000). Phospho-dependent association of neurofilament proteins with kinesin in situ. *Cell Motil. Cytoskeleton 45*, 249-262.

Yabe JT, Pimenta A, and Shea TB (1999). Kinesin-mediated transport of neurofilament protein oligomers in growing axons. *J. Cell Sci. 112 (Pt 21)*, 3799-3814.

Yang Y, Bauer C, Strasser G, Wollman R, Julien JP, and Fuchs E (1999). Integrators of the cytoskeleton that stabilize microtubules. *Cell 98*, 229-238.

Yin X, Crawford TO, Griffin JW, Tu P, Lee VM, Li C, Roder J, and Trapp BD (1998). Myelin-associated glycoprotein is a myelin signal that modulates the caliber of myelinated axons. *J. Neurosci. 18*, 1953-1962.

Yoshihara T, Yamamoto M, Hattori N, Misu K, Mori K, Koike H, and Sobue G (2002). Identification of novel sequence variants in the neurofilament-light gene in a Japanese population: analysis of Charcot-Marie-Tooth disease patients and normal individuals. J. Peripher. *Nerv. Syst. 7*, 221-224.

Zhang Z, Casey DM, Julien JP, and Xu Z (2002). Normal dendritic arborization in spinal motoneurons requires neurofilament subunit L. *J. Comp Neurol. 450*, 144-152.

Zheng YL, Li BS, Veeranna, and Pant HC (2003). Phosphorylation of the head domain of neurofilament protein (NF-M): a factor regulating topographic phosphorylation of NF-M tail domain KSP sites in neurons. *J. Biol. Chem. 278*, 24026-24032.

Zhu Q, Couillard-Després S, and Julien JP (1997). Delayed maturation of regenerating myelinated axons in mice lacking neurofilaments. *Exp. Neurol. 148*, 299-316.

Zuchner S, Vorgerd M, Sindern E, and Schroder JM (2004). The novel neurofilament light (NEFL) mutation Glu397Lys is associated with a clinically and morphologically heterogeneous type of Charcot-Marie-Tooth neuropathy. *Neuromuscul. Disord. 14*, 147-157.

In: New Research on Neurofilament Proteins
Editor: Roland K. Arlen, pp. 53-79

ISBN: 1-60021-396-0
© 2007 Nova Science Publishers, Inc.

Chapter III

Neurofilament Protein Partnership, Export, Transport, Phosphorylation and Neurodegeneration

Aidong Yuan[*]

New York University School of Medicine, Nathan Kline Institute,
Orangeburg, New York, U.S.A.

Abstract

Neurofilaments are about 10 nm diameter intermediate filaments of neurons which add rigidity and tensile strength to neurites (axons and dendrites) and determine neurite caliber. They are generally believed to be obligate heteropolymers composed of the neurofilament triplet proteins, designated NF-L, NF-M and NF-H for light, medium and heavy molecular weight subunits, respectively. Two other related intermediate proteins alpha-internexin and peripherin may be closely associated with neurofilament. Recent analyses of mice in which single or multiple neurofilament genes are deleted have defined the minimum structural requirements for efficient neurofilament export and transport and also showed the dissociation of neurofilament number in axons with the transport rate of neurofilament. After their synthesis, neurofilament proteins form proper partnership with one another and are exported into and transported throughout axons primarily in the form of heterooligomers or shot filaments where they are incorporated into or exchanged with a basically stationary neurofilament network. Neurofilament proteins distribute non-uniformly along axons with phosphorylation increasing proximally to distally. The gene deletion analyses and additional mutagenesis studies provide new evidences for us to reconsider the basic composition of neurofilament and

---

[*] Correspondence: Dr. Aidong Yuan, Center for Dementia Research, Nathan Kline Institute, New York University School of Medicine, 140 Old Orangeburg Road, Orangeburg, NY 10962. Email: yuan@nki.rfmh.org ; Fax: 845-398-5422.

how phosphorylation controls neurofilament transport. Surprisingly, deletion of the NF-H or its phosphorylated tail domain or NF-M phosphorylated tail domain does not alter neurofilament transport rate in vivo and, the NF-M tail domain rather than NF-H or its tail domain proved to be essential for determining the radial growth and conduction velocity of large myelinated axons. The observations that mutations of NF-L are a cause of Charcot-Marie-Tooth disease type 2E/1F, a neuronopathy associated with neurofilament accumulation and discovery of neurofilament inclusion disease, a form of frontal temporal dementia with hallmark lesions containing neurofilaments and alpha-internexin, support growing evidence that neurofilament may have a causative role in neurodegeneration. A model was proposed to explain the role of neurofilament in neurodegeneration, i.e., gene mutation or environmental damage on axons could cause loss of neurofilament protein proper partnership, inefficient export of neurofilament and accumulation of remaining partner in cell boy and proximal axons, resulting in age-related axonal atrophy or induce abnormal movement or retraction of axonal phosphorylated neurofilaments back into cell bodies or axonal spheroids, leading to progressive neurodegeneration and eventually cell death.

1. Introduction and History

In early 1800s, neuronal fibrous network was first discovered by neuro-anatomists. The neurofibrils of the fibrous network was visualized with the help of silver staining method developed in late 1800s and later identified by electron microscopy in mid 1900s as comprised of 10 nm diameter filaments and became known as neurofilaments (NF) (Schmitt, 1968). This apparent diameter is between that of actin filaments (5-6 nm) and microtubules (24-25 nm) and makes neurofilament a member of the intermiate filament (IF) proteins, which are expressed in different cell types. The triplet protein theory of neurofilament composing of three subunits light (NF-L), middle (NF-M) and heavy (NF-H) is widely accepted based on axonal transport studies and co-isolation in purified neurofilament fractions and co-assembly with neurofilament subunit (Hoffman and Lasek, 1975; Liem et al., 1978; Mori et al., 1979; Willard and Simon, 1981; Zackroff et al., 1982; Lasek et al., 1985). Recent advance in molecular biology of neurofilament and transgenic mice firmly establishes that neurofilament is essential for radial growth of axons which in turn controls conduction velocity, and defects in neurofilament partnership, export, transport, assembly and organization caused by gene mutations or environmental damage to neurons can be a primary cause or a key intermediate in the pathogenesis of both hereditary and sporadic neurodegenerative diseases.

2. Neurofilament Definition and in Relation to Other IF Proteins

Neurofilament means intermediate filaments of neurons, especially of axons. Its official definition from Unified Medical Language System at the National Library of Medicine is intermediate filaments of neurons which add rigidity, tensile strength, and possibly intracellular transport guidance to axons and dendrites. Neurofilament belongs to a member of the intermiate filament (IF) protein family, which are 10 nm filaments of fibrous system composed of chemically heterogeneous subunits and involved in mechanically integrating the various components of the cytoplasmic space. More than 70 different intermediate filament proteins have been identified (Hesse et al., 2001; Coulombe et al., 2004) and classified into five chemically distinct classes. Types I and II are keratin filaments of mammalian epithelial cells. Type III includes vimentin filaments found in mesenchymal cells; desmin filaments found in cells of muscles of all types; glial filaments found in all types of glial cells and peripherin filaments found mainly in PNS neurons. Type IV includes neurofilaments of neurons and α-internexin of mainly CNS neurons; nestin filaments found in CNS stem cells; synemin and syncoilin found in muscle cells and nonstandard type IV filensin and phakinin found in lens fiber cells. Type V is composed of lamin filaments of nuclear lamina (Table 1). Neurofilament genes, like all other cytoplasmic IF genes, are evolved from the duplication of the nuclear lamin genes.

Table 1. Family of IFP

Type	Protein	Molecular Weight	Cellular Location
I	Acidic keratins	40-60 kDa	Epithelial cells and their derivatives
II	Neutral or basic keratins	50-70 kDa	Epithelial cells and their derivatives
III	Vimentin	54-57 kDa	Many cells of mesenchymal origin (e.g. fibroblasts)
	Desmin	53 kDa	Muscle cells
	Glial fibrillary acidic protein	50-52 kDa	Glial cells (astrocytes and some Schwann cells)
	Peripherin	54-57 kDa	Mainly PNS neurons
IV	NF-L	61-70 kDa	Neurons
	NF-M	90-150 kDa	Neurons
	NF-H	110-200 kDa	Neurons
	α-internexin	58-66 kDa	CNS neurons
	Syncoilin	54-64 kDa	Mainly skeletal and cardiac muscle cells
	Nestin	200-240 kDa	CNS stem cells
	Synemin	150-230 kDa	Muscle cells
Nonstandard IV	Filensin	83-110 kDa	Lens fiber cells
	Phakinin	46-47 kDa	Lens fiber cells
V	Nuclear lamins A, B, C	60-78 kDa	Nuclear lamina

3. Neurofilament Composition

Neurofilaments are currently believed to be composed of three subunits, the so-called neurofilament triplet proteins based on their molecular weight. They are designated NF-L (light or low), NF-M (medium or middle) and NF-H (heavy or high), which are around 62 kDa, 102 kDa and 110 kDa, respectively, as calculated from their complete DNA sequence data in humans, whereas those in mouse are of 62 kDa, 96 kDa and 116 kDa, respectively. The migration of the triplet proteins is slow at around 68 kDa, 150 kDa and 200 kDa, respectively in SDS-PAGE due to their enriched negatively charged glutamic acid, phosphorylation and glycosylation. The official name of NF-L is "neurofilament, light polypeptide 68kDa". NEFL is the gene's official symbol. Other names used for NEFL gene or gene products include neurofilament triplet L protein, NF68 or NFL. The official name of NF-M is "neurofilament 3 (150kDa medium)". NEF3 is the gene's official symbol. Other names used for NEF3 gene or gene products include neurofilament triplet M protein, neurofilament medium polypeptide, NFM and NEFM. The official name of NF-H is "neurofilament, heavy polypeptide 200kDa". NEFH is the gene's official symbol. Other names used for NEFH gene or gene products include neurofilament triplet H protein, NFH and KIAA0845. Two other related intermediate proteins alpha-internexin and peripherin may be closely associated with neurofilament. Recent analyses of mice in which single or multiple neurofilament genes are deleted have showed that α-internexin qualifies more as a neurofilament subunit than NF-H in terms of axonal transport, because deleting α-internexin in the absence of NF-L reduces NF-M transport to undetectable levels whereas deleting NF-H in the absence of NF-L has no apparent effect on NF-M transport (Yuan et al., 2003). Besides the triplet proteins and α-internexin, other minor unidentified proteins could also be intimately associated with the neurofilament and contribute its structure and function.

4. Neurofilament Assembly and Structure

Neurofilament proteins, like all other IF proteins, have a central α-helical rod domain of approximately 310 amino acid (350 amino acid in the nuclear lamins). This central rod domain is flanked by amino-terminal head domain and carboxy-terminal tail domain, which vary among the different IF proteins (Figure 1). The central α-helical rod domain of NF-L can form a parallel coiled-coil dimer with that of NF-M. A pair of these parallel NF-L / NF-M heterodimers then associates in an antiparallel fashion to form a staggered tetramer containing 2 NF-L and 2 NF-M, which can assemble end to end to form protofilaments (2 - 3 nm thick) containing 4 NF-L and 4 NF-M. Approximately eight protofilaments (32 NF-L and 32 NF-M) wind around each other to form the final neurofilament 10 nm in diameter (Figure 2). Because assembled from anti-parallel tetramers, neurofilaments are apolar in contrast to actin filaments and microtubules, which are polar structures. Unlike polar actin filaments and microtubules whose ends grow at different rates, new neurofilament proteins add along the neurofilament length as well as the neurofilament ends and there are no known examples of motor proteins that move along neurofilaments. It is believed that neurofilaments are obligate heteropolymers requiring NF-L and substoichiometric amounts of NF-M and or NF-H (Lee et al., 1993). Recent analyses of mice in which single or multiple neurofilament genes are deleted have showed, however, NF-M and α-internexin can form neurofilaments in the absence of both NF-L and NF-H (Yuan et al., 2003).

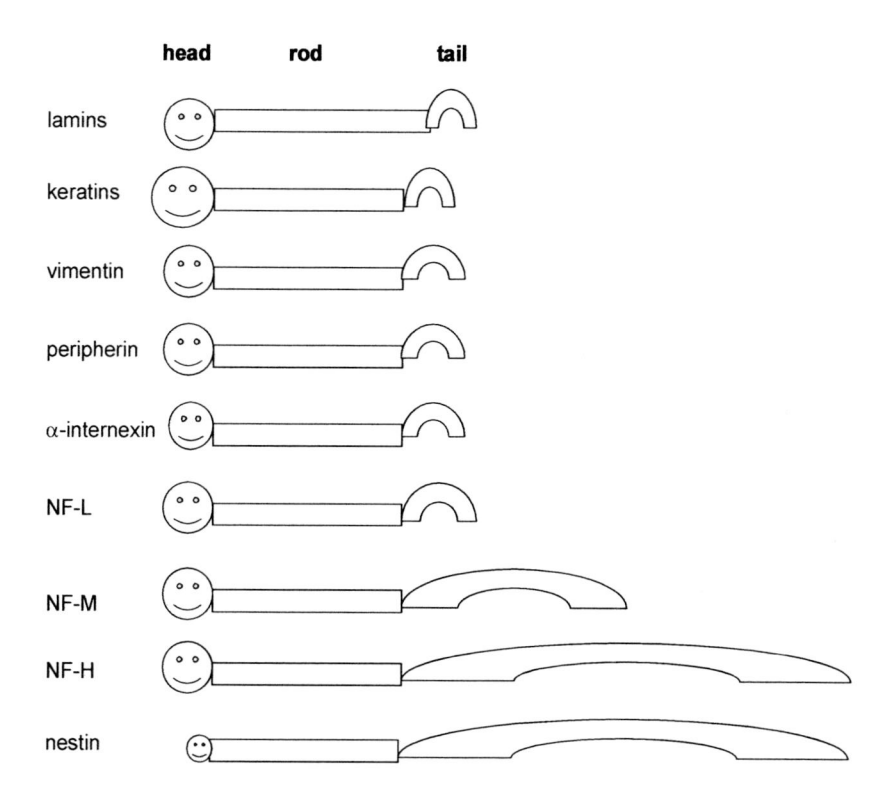

Figure 1. The domain organization of IFP. IFP consist of 3 domains, an amino-terminal globular head, conserved alpha-helical rod and a carboxy-terminal tail. Both the head and tail domains vary greatly in sequence and length in different IFP monomers.

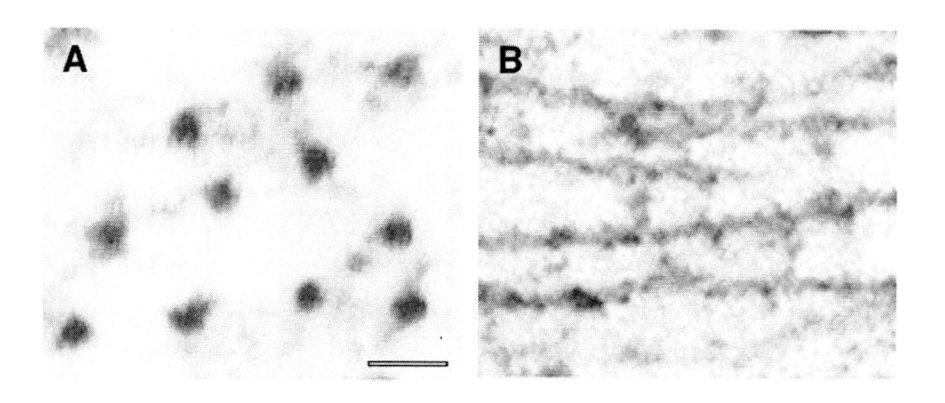

Figiure 2. Electron micrographs of neurofilament in a nerve cell axon. (A) cross-section; (B) longitudinal section. Scale bar, 20 nm.

5. Neurofilament Expression and Function

Neurofilament proteins exist in all vertebrates, and are believed to have originated by gene duplication from an ancestral NF-M-like protein (Jacobs et al., 1995). The most primitive living vertebrate lamprey has only a single NF-M-like subunit (NF-180) (Hall et al., 2000), but more recent studies suggest that this subunit may need another previously unrecognized IF protein for filament assembly (Jin et al., 2005). In tiger salamander olfactory axons NF-M is abundant but NF-L and NF-H subunits are present at trace levels, resulting in rare neurofilament formation (Burton and Wentz, 1992). In developing spinal cord neurons of *Xenopus laevis*, rat cerebral cortex and mouse myenteric neurons, NF-M and α-internexin (or α-internexin-like *Xenopus* protein) emerge simultaneously before the appearance of NF-L and NF-H subunits indicating a close relationship between NF-M and α-internexin (Giasson and Mushynski, 1997; Faussone-Pellegrini et al., 1999; Undamatla and Szaro, 2001). The human NF-L and NF-M genes are closely located at chromosome 8p21 (about 30 kb apart) while NF-H and α-internexin genes are located at chromosome 22q12.2 and chromosome 10q24.33, respectively. There is a delayed appearance of NF-H during early development due to its low levels of expression compared with all other neurofilament proteins (Shaw and Weber, 1982; Willard and Simon, 1983; Pachter and Liem, 1984; Carden et al., 1987; Sanchez et al., 2000). The most important function of neurofilament is the establishment of axonal caliber, which is a key determinant of conduction velocity, especially in large myelinated nerve fibers (Hursh, 1939; Friede and Samorajski, 1970; Waxman, 1980; Nixon and Logvinenko, 1986). Evidence supporting this derives first from the linear relationship between neurofilament content and axonal diameter (Nixon and Logvinenko, 1986), and the correlation of reduction in axonal caliber after axotomy with a selective reduction of neurofilament gene expression (Hoffman et al., 1987). Later studies of naturally occurring mutant Japanese quail (Yamasaki et al., 1991; Yamasaki et al., 1992; Ohara et al., 1993; Sakaguchi et al., 1993) and transgenic mice lacking axonal neurofilaments prove unequivocally this fundamental role of neurofilament (Eyer and Peterson, 1994; Zhu et al., 1997; Elder et al., 1998a; Jacomy et al., 1999). Recent advances in gene knock-in approach reveal it is NF-M tail domain (Garcia et al., 2003; Rao et al., 2003) rather than NF-H (Elder et al., 1998b; Rao et al., 1998; Zhu et al., 1998) or its tail domain (Rao et al., 2002) that is essential for determining the radial growth and conduction velocity of large myelinated axons. Besides its structural role, neurofilament may be involved in other functions such as regulation of cell surface expression and desensitization of D1 dopamine receptor through its subunit NF-M ail domain (Kim et al., 2002).

6. Neurofilament Protein Partnership, Export and Transport

Because axon has little protein synthesis machinery, neurofilament proteins are synthesized in the cell bodies and exported into and transported through axons by a process called slow axonal transport. Neurofilament assembly requires at minimum the polymerization of NF-L with either NF-M or NF-H subunits, but requirements for their axonal transport have long been controversial. A long-standing debate has centered on the form of the transported neurofilament proteins (Mills et al., 1996; Baas and Brown, 1997; Bray, 1997; Hirokawa et al., 1997; Yuan, 1997; Nixon, 1998; Galbraith et al., 1999; Yabe et al., 1999; Brown, 2000; Prahlad et al., 2000; Terada and Hirokawa, 2000; Yuan et al., 2000; Shah and Cleveland, 2002; Yan and Brown, 2005). By one view, neurofilament proteins are assembled in the cell body and transported into axons as fully assembled neurofilaments, i.e., the polymer transport model (Baas and Brown, 1997). Supporting this view is the direct observation of a subpopulation of GFP-tagged neurofilaments moving bidirectionally in growing axons of cultured embryonic neurons (Roy et al., 2000; Wang et al., 2000; Wang and Brown, 2001; Ackerley et al., 2003; Uchida and Brown, 2004). Alternatively, neurofilament proteins have been proposed to move in subunit form (monomer or oligomer) and exchange with those in the neurofilaments that form a stable network in the axon, i.e., the subunit transport model (Hirokawa et al., 1997). In support of this model, virally expressed NF-M were transported at slow rates without NF-L and NF-H into axons in a transgenic mouse model that expresses a mutant NF-H linked to beta-galactosidase, which aggregates the neurofilaments within the perikaryon and substantially reduces the numbers of neurofilaments in the axons (Terada et al., 1996). The differential dynamics of NF-L and NF-H in axons *in vivo* have also been interpreted as support for subunit movement (Takeda et al., 1994). Recently, studies on cultured neuronal cell lines or chick DRG neurons have also reported the axonal transport of GFP-tagged neurofilament protein and peripherin in a nonfilamentous form (Yabe et al., 1999; Helfand et al., 2003; Theiss et al., 2005). Conflicting studies on the form of transported neurofilament proteins is also reported in squid axons (Galbraith et al., 1999; Prahlad et al., 2000).

Pulse-labeling experiments are typically used for measuring axonal transport rates of neurofilament. Radiolabeled protein precursors (e.g. 35S-methionine) are injected into the fluid bathing nerve cell bodies in vivo (e.g. the vicinity of retinal ganglion cells and anterior horn motor neurons of lumbar spinal cord). The radioactive amino acids became incorporated into neurofilament proteins that are then transported along the axon of the cell. After different times ranging from hours to months, the nerves are harvested and cut into segments, and each segment is analyzed using SDS-PAGE and autoradiography to identify the transport rates of the radioactive neurofilament proteins. As shown in Figure 3, radioactive NF-M at 5 hours after injection is mainly present in segment 1, i.e., 1 mm from the eye with little radiolabeled NF-M moved into segment 2 (i.e., 2 mm from the eye). At 1 day post-injection, some radioactive NF-M protein (the front wave) is present at 4 mm away from the eye while the most heavily labeled NF-M band is still in segment 1. The transport rate of front radioactive NF-M is about 4 mm/day. At 3 days post-injection, some radioactive NF-M protein is present at 7 mm away from the eye while the most heavily labeled NF-M band is still in segment 1. At 7, 14, and 21 days post-injection, the most heavily labeled NF-M band moves into segment 2, 3 and 4, respectively. Therefore, the peak rate of radioactive NF-M is about 1 mm in 7 days, i.e., 0.14 mm/day.

Recently, we directly addressed in mice whether or not neurofilament subunits are capable of independent translocation in pulse-labeling experiments. Using a gene deletion approach, we generated mice containing only NF-L or NF-M. In vivo pulse radiolabeling analyses in retinal ganglion cell neurons revealed that NF-L alone is incapable of efficient export, whereas nearly half of the normal level of NF-M is exported into and transported along optic axons in the absence of both NF-L and NF-H subunits. The number of intermediate filaments in optic axons of NF-H and NF-L double knockout (HL-DKO) mice is less than 10% of the number of neurofilaments in wild-type mouse axons (A. Yuan and R. A. Nixon, unpublished data) and unlikely to account for the high proportion of NF-M that is still transported in HL-DKO mice. In these axons, more likely the transported NF-M in HL-DKO mice is in the form of both hetero-oligomers and short hetero-polymers. Deleting

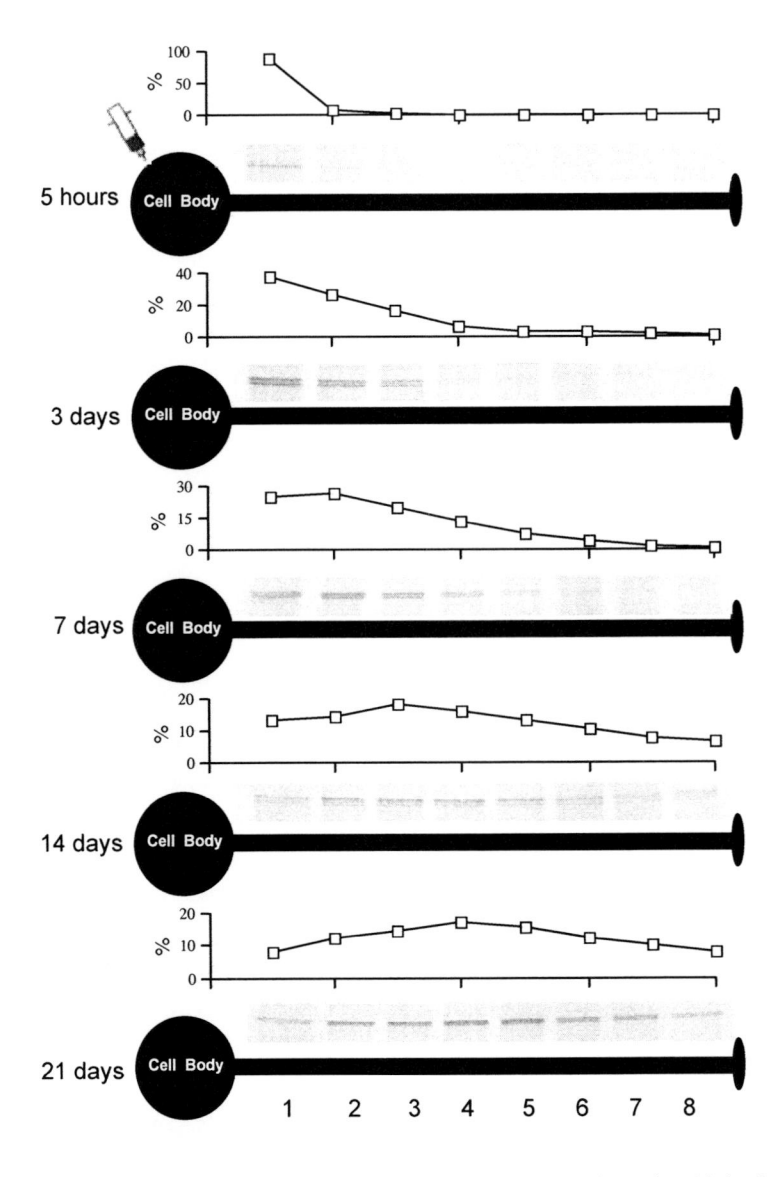

Figure 3. NF transport assay. The retinal ganglion cells are radiolabeled by intravitreal injection of 35S-methionine. After different times (from 5 hours to 21 days), the optic pathways are harvested and cut into eight consecutive 1mm segments. The segments can be used to evaluate rate of NF transport (represented by NF-M) using SDS-PAGE and autoradiography.

α-internexin in the absence of NF-L reduces NF-M transport to undetectable level with small amount of steady state NF-M and NF-H proteins present in optic axons. These results suggest that although neurofilament subunits can move into axon, their efficient axonal transport units are hetero-oligomers and short hetero-polymers. NF-M can partner with NF-L and α-internexin to support slow transport and possibly other functions of neuronal intermediate filaments. These findings define important structural requirements for the interactions of neurofilament subunits with the putative motors that mediate slow transport (Yuan et al., 2003). More recently, it has been reported that neurofilament may interact with cytoplasmic dynein through the specific binding of NF-M rod domain (122-428aa) directly to dynein intermediate chain (Wagner et al., 2004).

Based on recent advances in molecular genetics, I think the polymer and subunit models are not mutually exclusive, at least for neurofilament proteins. Transport of fully assembly long neurofilament or neurofilament networks would use more energy than that of pre-assembled oligomers and short filaments, and may require multiple molecules of motors and multiple microtubule tracks, as in the case of mitochondria transport, which is usually more than 0.5 μm in size. Transport of neurofilament protein monomers is also less energy-efficient than moving pre-assembled hetero-oligomers or short neurofilament. The combined model appears particularly intriguing since this mode of transport would be not only consistent with energy efficiency during evolution but also provide a supply of pre-assembled precursors for repair of the stable neurofilament networks in the event of injury. The combined model predicts that in a mature axon, neurofilament networks are basically stationary but dynamic, which are constantly renewed by exported moving precursors, i.e., pre-assembled hetero-oligomers and short hetero-polymers (Figure 4).

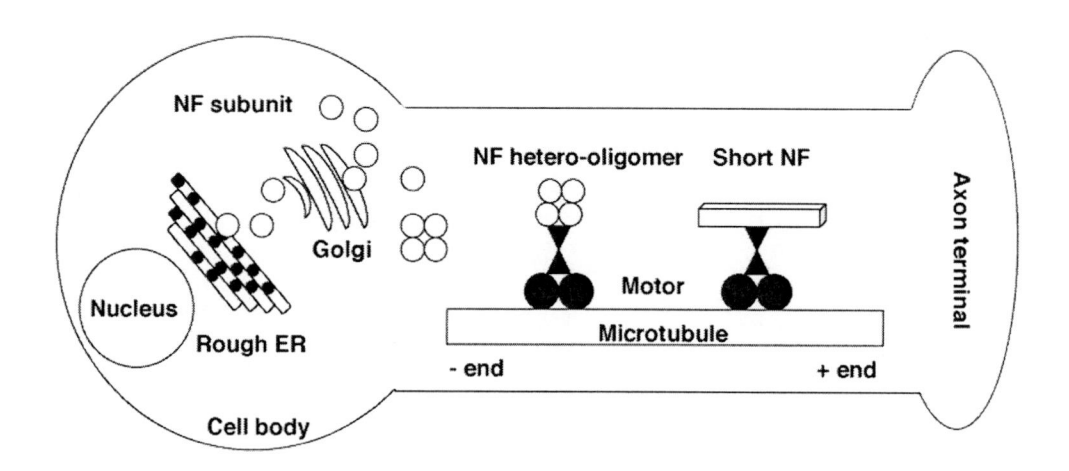

Figure 4. A model for the axonal transport of NFP. After synthesis, NFP monomers primarily form hetero-oligomers or short filament in cell body for efficient transport into axon and integration into NF network. A small amount of NFP monomers or homo-oligomers could be much less efficiently transported into axons.

7. Neurofilament Protein Phosphorylation

Neurofilament proteins are phospho-proteins and phosphorylated at both ends of the subunits (Nixon et al., 1982; Sternberger and Sternberger, 1983; Lee et al., 1987; Sihag and Nixon, 1990). They are the most extensively phosphorylated proteins in neurons (Julien and Mushynski, 1982, 1983; Carden et al., 1985). After their synthesis in cell body, neurofilament proteins are transported into and throughout axons and are more and more phosphorylated during transport toward axonal terminals, especially the Lys-Ser-Pro (KSP) repeat domains of NF-M and NF-H C-terminals (Nixon et al., 1987). For example, rat NF-L, NF-M and NF-H have 3, 9, and 22 phosphates, respectively (Julien and Mushynski, 1982). Both in culture (Bennett et al., 1984; Brown, 1998) and in vivo (Szaro et al., 1989; Archer et al., 1994), steady state neurofilament protein phosphorylation increases proximally to distally along axons (Figure 5), which is responsive to neuronal stress and regulated by various non-proline directed and proline directed kinases and several phosphatases (Grant and Pant, 2000). Introduction of a mutant NF-L transgene in which Ser55 is mutated to Asp so as to mimic permanent phosphorylation in transgenic mice causes formation of neurofilament aggregates in brain neuronal cell bodies (Gibb et al., 1998), suggesting a role of NF-L S35 phosphorylation in filament organization. Neurofilament protein phosphorylation is believed to influence axonal caliber and nerve conduction velocity. Unexpectedly, it is not NF-H tail domain with its multiple KSPs (Rao et al., 2002) but NF-M tail domain with its phosphorylation sites that is essential for the radial growth of axons (Rao et al., 2003) given the fact that phosphorylated NF-H contains much more phosphates (50 mol phosphate) than phosphorylated NF-M (15 mol phosphate) (Jones and Williams, 1982; Julien and Mushynski, 1982; Geisler et al., 1987; Goldstein et al., 1987; Lee et al., 1988).

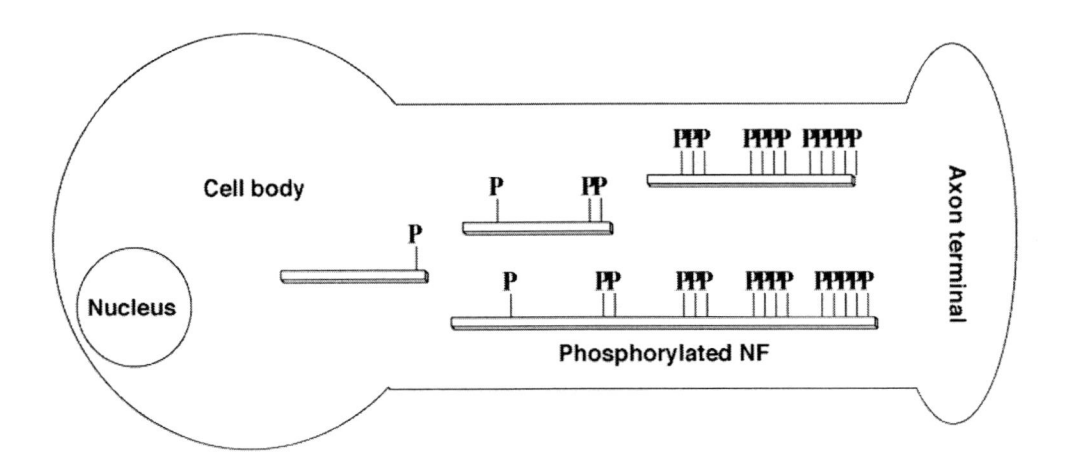

Figure 5. NFP phosphorylastion increases along axon in a proximal-to-distal manner. NFP are heavily phosphorylated at distal, intermediately at middle and lightly at proximal part of axon.

8. Disassociation of Neurofilament Numbers from its Transport Rate

Changes of neurofilament transport have been implicated during development and under certain pathological conditions. NF transport is gradually retarded during maturation and aging (McQuarrie et al., 1989). Neurofilament accumulations in cell bodies and at proximal axons are a prominent pathological feature of several neurofibrillary degenerative diseases such as neurofilament inclusion disease, amyotrophic lateral sclerosis (ALS), Parkinson's disease, Alzheimers's disease, dementia with Lewy bodies, progressive supranuclear palsy, Charcot-Marie-Tooth disease, diabetic neuropathy, and giant axonal neuropathy (Hirano, 1991; Trojanowski et al., 1993; Lee and Cleveland, 1996; Julien and Mushynski, 1998; Julien, 1999; Gotow, 2000; Perrone Capano et al., 2001; Ruiz-Ederra and Vecino, 2001; Fernyhough and Schmidt, 2002; Rao and Nixon, 2002; Al-Chalabi and Miller, 2003; Rao and Nixon, 2003; Bruijn et al., 2004; Cairns et al., 2004; Lariviere and Julien, 2004; Liu et al., 2004; Petzold, 2005; Andersen, 2006; Bruijn and Cudkowicz, 2006). Decreased rate of NF transport has been reported in ALS mouse model expressing mutant SOD-1 (Zhang et al., 1997; Williamson and Cleveland, 1999) or human (Collard et al., 1995) or mouse NF-H (Marszalek et al., 1996), and also in sciatic nerves of hypothyroid rats (Sidenius et al., 1987) or optic nerves of hypothyroid mice (Stein et al., 1991). Increased rate of NF transport has been reported in an experimental rat model of giant axonal neuropathy (Monaco et al., 1985; Monaco et al., 1990). It is not clear, however, what the precise mechanisms are by which neurofilaments accumulate. It was previously hypothesized that neurofilament number in axons is controlled by the axonal transport rate of neurofilaments (Hoffman et al., 1984; Hoffman et al., 1985) and NF-H subunit modulates axonal diameter by selectively slowing neurofilament transport rate (Marszalek et al., 1996). It is also suggested that slowed transport rate of neurofilaments causes perikaryal and proximal axonal accumulation of neurofilaments (Collard et al., 1995; Marszalek et al., 1996; Zhang et al., 1997; Williamson and Cleveland, 1999) and accelerated transported rate of neurofilaments leads to giant axonal neuropathy (Monaco et al., 1985; Monaco et al., 1990).

Recently, we generated mice lacking both NF-L and NF-H using a gene deletion approach. Electron microscopy studies showed that the optic axons of these HL-DKO mice have greatly reduced caliber and less than 10% of the number of neurofilaments in wild-type mouse axons (A. Yuan, A. Kumar and R. A. Nixon, unpublished data) whereas in vivo pulse radiolabeling analyses in retinal ganglion cell neurons revealed that nearly half of the normal level of NF-M is transported at normal SCa rate along optic axons in the absence of both NF-L and NF-H subunits. These data prove that neurofilament number is dissociated from its transport rate, and axonal caliber is controlled by neurofilament number, not its transport rate as they are currently defined (Yuan et al., 2003).

9. Neurofilament Transport Rate does not Depend on its Phosphorylated Tail Domains

It has been proposed that NF-H and NF-M tail domain phosphorylation is a key determinant of neurofilament transport rate (Lewis and Nixon, 1988; Nixon et al., 1994; Yabe et al., 2001; Ackerley et al., 2003; Chan et al., 2004) and incorporation of the NF-H subunit slows the rate of neurofilament transport (Willard and Simon, 1983). Consistent with this hypothesis is the finding that in vivo deletion of NF-H increases (Zhu et al., 1998) while NF-H overexpression in mice decreases neurofilament transport rate in sciatic nerves (Collard et al., 1995; Marszalek et al., 1996). Moreover, less phosphorylated species of NF-H are transported at a much faster rate than more phosphorylated forms (Jung et al., 2000b; Jung et al., 2000a) and reducing the phosphorylation state of some of the KSP repeats of NF-H molecule, in cultured cortical neurons overexpressing this subunit, accelerates neurofilament transport rate (Ackerley et al., 2003). Thus, repetitive phosphorylation of the tails of NF-H and NF-M subunits has seemed a reasonable mechanism to regulate neurofilament transport rate. However, our recent transport studies in optic nerves of NF-H null mice do not exhibit altered rates of neurofilament transport (Rao et al., 2002). Moreover, NF-H tail domain deletion in mice resulting in complete loss of all potential tail domain phosphorylation sites does not affect neurofilament transport rate although hyperphosphorylation of NF-M is observed which could have compensated for modulating neurofilament transport rate (Rao et al., 2002; Yuan et al., 2006). Further transport studies in NF-M tailless mice lacking all potential tail domain phosphorylation sites does not alter neurofilament transport rate, either (Rao et al., 2003). It is possible that the presence of the phosphorylated tail on either the NF-H or NF-M is sufficient to mask effects on neurofilament transport of deleting the tail on the other subunit. This possibility is now tested in mice in which both NF-H and NF-M tails are eliminated and ruled out because deletion of both NF-H and NF-M tails do not alter neurofilament transport rate (Rao et al., 2004). Altogether, these results suggest that neurofilament transport rate is independent of the phosphorylated tail domains of NF-H and NF-M molecules even if phosphorylation could play a role in the modulation of neurofilament transport rate, which may involve the phosphorylation and dephosphorylation of transport motor, track or associated proteins of neurofilament.

10. Neurofilament Protein Inefficient Export, Misaccumulation and Axonal Atrophy

Axonal transport of pre-assembled neurofilament precursors (hetero-oligomer or short filament) may be the consequence of neurons trying to maximize its energy use efficiency during evolution. Any genetic and environmental damage causing inappropriate partnership or complete loss of partners before neurofilament proteins leave for axons could potentially give rise to inefficient export of neurofilament protein heteromers into axon, misaccumulation of remaining partners in cell body and ultimately lead to age-related axonal atrophy due to lack of axonal neurofilament network or collapse of established neurofilament network. The typical example is that NF-L protein is not efficiently exported into axon in the absence of its binding partners NF-M and NF-H (Figure 6), resulting in misaccumulation of lone NF-L in nerve cell body [(Jacomy et al., 1999; Yuan et al., 2003); A. Yuan and RA. Nixon, unpublished data)]. Another example is that NF-M and NF-H proteins are not efficiently exported into axon in the absence of their binding partner NF-L, leading to misaccumulation of NF-M and NF-H proteins in the perikaryon in neurofilament deficient Japanese quail (quv) that lacks neurofilament as the result of a premature translation terminator in the NF-L gene (Ohara et al., 1993; Toyoshima et al., 2000).

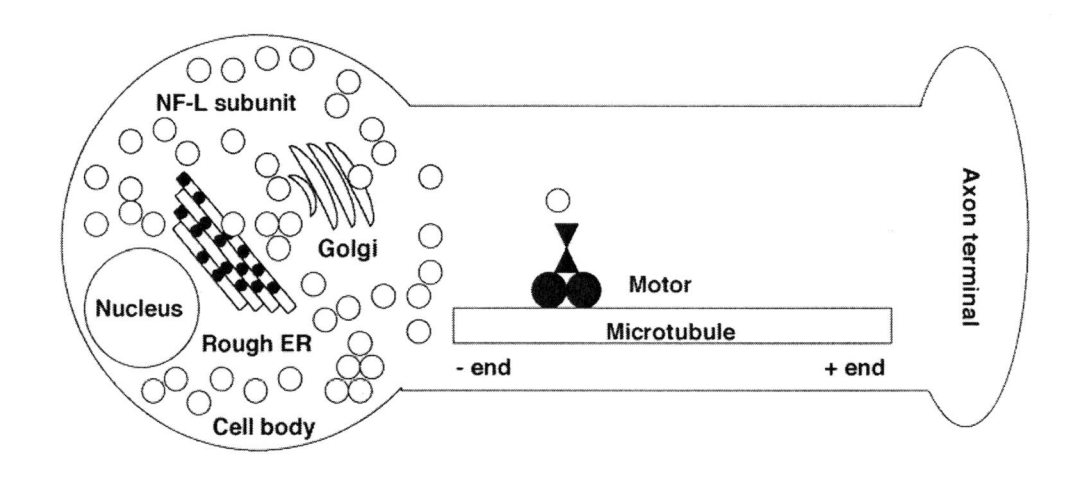

Figure 6. NF-L monomers misaccumulate in cell body. In the absence of its favorite partner NF-M and less favorite NF-H, NF-L monomers are not efficiently transported into axon and thus misaccumulate in perikaryon, resulting in swollen of cell body and age-related axonal atrophy.

Inefficient export and misaccumulation of neurofilament proteins can have neurological consequences as demonstrated by reduced conduction velocity (Sakaguchi et al., 1993), generalized ataxia and quivering in quv (Yamasaki et al., 1991) and age-related hind limb paralysis in mice (Elder et al., 1999). The observations that NF-L mutations cause Charcot-Marie-Tooth disease 2E and 1F (Mersiyanova et al., 2000; De Jonghe et al., 2001; Georgiou et al., 2002; Yoshihara et al., 2002; Jordanova et al., 2003; Zuchner et al., 2004), establish unequivocally a pathogenic role of neurofilament in neurodegenerative disease. These NF-L pathogenic mutations impair proper partnership of neurofilament proteins for efficient export from cell bodies and subsequent perikaryal misaccumulation (Brownlees et al., 2002; Perez-Olle et al., 2002; Perez-Olle et al., 2005; Sasaki et al., 2006), which could progressively disrupt the global balance of protein folding quality control, resulting in neurodegeneration (Gidalevitz et al., 2006).

11. Retraction of Phosphorylated Neurofilament Network and Neurodegeneration

Axonal transport and subsequent establishment of increasing proximal-distal gradient of phosphorylated neurofilament network along axon is essential for normal nerve function. This neurofilament organization is gradually built-up during development, well maintained during adulthood and little impaired during normal aging process. However, genetic and environmental damages directly or indirectly to this dynamic network could cause its abrupt breakage or progressive collapse leading to abnormal movement of phosphorylated neurofilament network or retraction back into cell body or axonal spheroids during neurodegeneration (Figure 7) (Sternberger and Sternberger, 1983; Drager and Hofbauer, 1984; Strong et al., 2001).

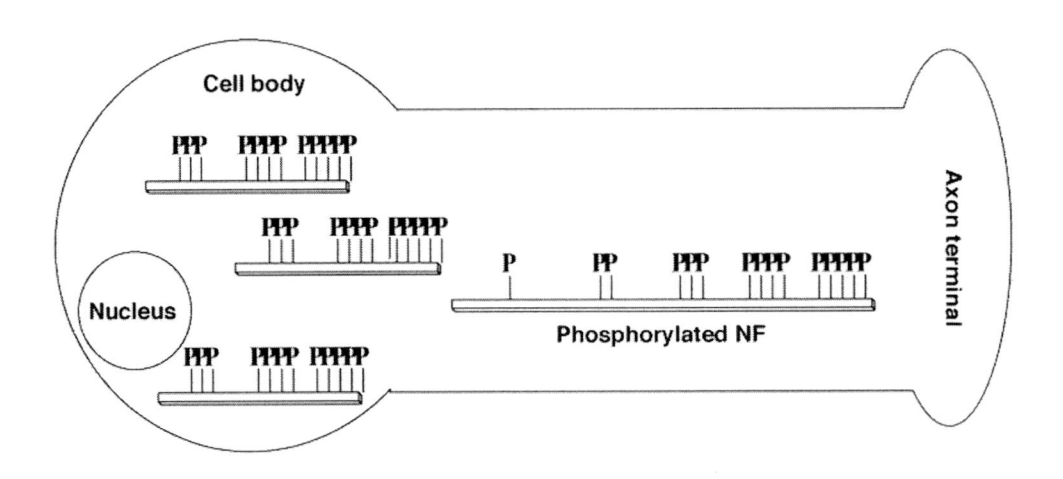

Figure 7. Phosphorylated NF misaccumulates in cell body or axonal spheroids. Highly phosphorylated NF organization in axon is vulnerable to both genetic mutations and environmental damage, causing retraction of phosphorylated NF back into cell body or axonal spheroids.

Although neurofilament gene mutations alone are sufficient to cause neurodegenerative disease and kill neurons (Charcot-Marie-Tooth disease 2E/1F - CMT2E/1F) in the absence of other initiating factors, prominent neurofilament pathology occurs in a wide variety of other neurodegenerative diseases such as neurofilament inclusion disease, amyotrophic lateral sclerosis (ALS), Parkinson's disease, Alzheimers's disease, dementia with Lewy bodies, progressive supranuclear palsy, diabetic neuropathy, giant axonal neuropathy and some other forms of Charcot-Marie-Tooth disease (Perrone Capano et al., 2001; Al-Chalabi and Miller, 2003; Liu et al., 2004; Petzold, 2005). This leads to the hypothesis that neurofilament may play a role in a final common pathway or act as a key intermediate in the pathogenesis of neuronal degeneration and death, which can be initiated by other pathogenic triggers. Besides the pathogenic mutations of the NF-L gene in CMT2E/1F, variations of the NF-H gene have been associated with increased risk of developing ALS and a point mutation has been reported in the rod domain of the NF-M gene in an individual with Parkinson's disease

(Lavedan et al., 2002). Although mutations of neurofilament protein cause CMT2E/1F or increase the risk of developing ALS, un-mutated neurofilament protein and α-internexin can also be the signature lesions in a recently identified dementia characterized by inclusions of neurofilaments and α-internexin which do not contain beta-amyloid and prion deposits nor tau or alpha-synuclein protein aggregates (Cairns et al., 2003; Josephs et al., 2003; Cairns et al., 2004; Momeni et al., 2005). Peripherin, a type III intermediate filament protein, is present mainly in neurons of the mammalian peripheral nervous system and its mutations are associated with an increased susceptibility to development of ALS (Gros-Louis et al., 2004; Leung et al., 2004). In addition to variations of neuronal intermediate filament proteins, other gene mutations can also lead to disruption of neurofilament organization including HSPB1 (heat shock 27kDa protein 1) mutations in CMT2F (Evgrafov et al., 2004), gigaxonin mutations in giant axonal neuropathy (Bomont et al., 2000) and SOD1 mutations in ALS (Rosen et al., 1993) (Table 2). Interestingly, HSPB1 and SOD1 are both carried in slow axonal transport (Yuan, 1997; Borchelt et al., 1998) where they may play a role in neurofilament transport and assembly (Ackerley et al., 2006) or maintaining the stability of neurofilament network (Menzies et al., 2002). Gigaxonin is a member of the cytoskeletal BTB-kelch-repeat superfamily and can bind to MAP1B and control its degradation (Ding et al., 2002; Allen et al., 2005). Mutation of this gigaxonin gene causes excess neurofilament in the axons, which do not transmit action potential properly and eventually deteriorate, resulting in dysfunctions of both motor and sensory nervous system.

Table 2. NF-related disease mutations

Gene	Mutations or abnormalities	Disease type
NF-L	Missense mutations or codon deletions	Charcot-Marie-Tooth disease 2E and 1F
	Nonsense mutation	Quiver - NF deficient Japanese quail (Quv)
NF-M	Missense mutations	Parkinson's disease
NF-H	Codon deletions or insertions	Amyotrophic lateral sclerosis
Neurofilament triplet and α-internexin	NF aggregation	Dementia with NF inclusions
Peripherin	Insertions or deletions disrupting NF assembly	Amyotrophic lateral sclerosis
HSPB1	Missense mutations disrupting NF assembly	Charcot-Marie-Tooth disease 2F
Gigaxonin	Missense, nonsense, deletion and insertion mutations disrupting NF assembly	Giant axonal neuropathy
SOD1	Missense mutations disrupting NF organization	Amyotrophic lateral sclerosis

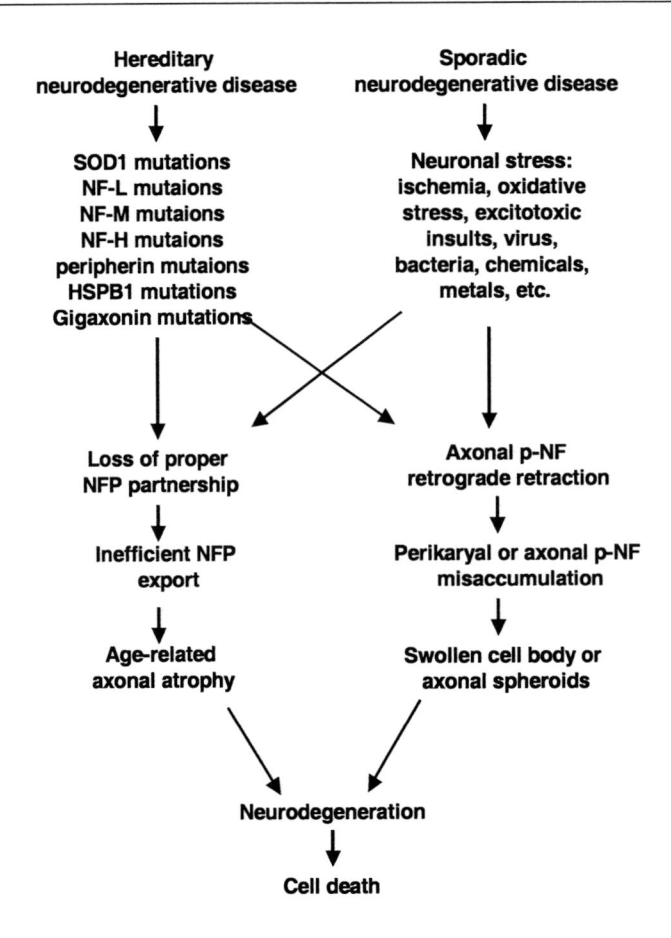

Figure 8. A model for disrupted NFP export, transport and assembly in hereditary and sporadic neurodegenerative diseases. Inherited NF-L mutations, such as in CMT2E/1F, lead to loss of proper NFP partnership for their efficient export from cell bodies into axon and integration into stable NF network, resulting in age-related axonal atrophy and neurodegeneration. Other forms of neuronal stress, such as ischemic injury or toxins may induce retraction of phosphorylated NF back into cell body or axonal spheroids, leading to malfunction of cell body, axon and synapse.

12. Conclusion

Neurofilament is intermediate filament of neuron, and its proper export from cell bodies and transport throughout axons and establishment of phosphorylation gradient is essential for normal nerve function. Although it is generally believed to be obligate heteropolymers composed of the neurofilament triplet proteins (NF-L, NF-M and NF-H), its definition is evolving and may well contain other subunit or subunits in vivo. Efficient export and transport unit for neurofilament proteins is heteromers including possibly hetero-dimer, hetero-oligomer and short filament. Axon caliber is controlled by neurofilament number, which is disassociated from its transport rate. Deletions of the NF-H or its phosphorylated tail domain or NF-M phosphorylated tail domain or both phosphorylated tail domains of NF-H and NF-M do not alter neurofilament transport rate in vivo. The NF-M tail domain rather than

NF-H or its tail domain proved to be essential for determining the radial growth and conduction velocity of large myelinated axons. The observations that NF-L mutations cause Charcot-Marie-Tooth disease 2E and 1F, establish unequivocally a pathogenic role of neurofilament in neurodegeneration. Genetic and environmental damages could impair proper partnership of neurofilament proteins for their export from cell bodies and transport throughout axons and subsequent perikaryal misaccumulation, or abnormal movement and retraction of axonal phosphorylated neurofilaments back into cell bodies or axonal spheroids, leading to progressive neurodegeneration (Figure 8). Future challenges lie ahead for translating this knowledge into preventions and therapies capable of repairing the nervous system and improving neuronal functions in neurodegenerative diseases.

13. Acknowledgements

I would like to thank my mentor Dr. Ralph Nixon for his continuing encouragement and full support for the neurofilament work (grant AG05604). My thanks also go to Drs. Jean-Pierre Julien, Mala Rao, Asok Kumar, Yuanxin Chen, Veeranna, Panaiyur Mohan, Takahiro Sasaki, Ron Liem and Alan Peterson for their enormous help with the neurofilament studies.

Note Added in Proof

Since this review was written alpha-internexin has now been shown to be a fourth subunit of neurofilament in the adult CNS, providing a basis for its close relationship with neurofilament in CNS diseases associated with neurofilament accumulation. This reference is provided below.

Yuan A, Rao MV, Sasaki T, Chen Y, Kumar A, Veeranna, Liem RK, Eyer J, Peterson AC, Julien JP, Nixon RA (2006) Alpha-internexin is structurally and functionally associated with the neurofilament triplet proteins in the mature CNS. *J Neurosci* 26:10006-10019.

14. References

Ackerley S, James PA, Kalli A, French S, Davies KE, Talbot K (2006) A mutation in the small heat-shock protein HSPB1 leading to distal hereditary motor neuronopathy disrupts neurofilament assembly and the axonal transport of specific cellular cargoes. *Hum Mol Genet* 15:347-354.

Ackerley S, Thornhill P, Grierson AJ, Brownlees J, Anderton BH, Leigh PN, Shaw CE, Miller CC (2003) Neurofilament heavy chain side arm phosphorylation regulates axonal transport of neurofilaments. *J Cell Biol* 161:489-495.

Al-Chalabi A, Miller CC (2003) Neurofilaments and neurological disease. *Bioessays* 25:346-355.

Allen E, Ding J, Wang W, Pramanik S, Chou J, Yau V, Yang Y (2005) Gigaxonin-controlled degradation of MAP1B light chain is critical to neuronal survival. *Nature* 438:224-228.

Andersen PM (2006) Amyotrophic lateral sclerosis associated with mutations in the CuZn superoxide dismutase gene. *Curr Neurol Neurosci Rep* 6:37-46.

Archer DR, Watson DF, Griffin JW (1994) Phosphorylation-dependent immunoreactivity of neurofilaments and the rate of slow axonal transport in the central and peripheral axons of the rat dorsal root ganglion. *J Neurochem* 62:1119-1125.

Baas PW, Brown A (1997) Slow axonal transport: the polymer transport model. *Trends Cell Biol* 7:380-384.

Bennett GS, Tapscott SJ, DiLullo C, Holtzer H (1984) Differential binding of antibodies against the neurofilament triplet proteins in different avian neurons. *Brain Res* 304:291-302.

Bomont P, Cavalier L, Blondeau F, Ben Hamida C, Belal S, Tazir M, Demir E, Topaloglu H, Korinthenberg R, Tuysuz B, Landrieu P, Hentati F, Koenig M (2000) The gene encoding gigaxonin, a new member of the cytoskeletal BTB/kelch repeat family, is mutated in giant axonal neuropathy. *Nat Genet* 26:370-374.

Borchelt DR, Wong PC, Becher MW, Pardo CA, Lee MK, Xu ZS, Thinakaran G, Jenkins NA, Copeland NG, Sisodia SS, Cleveland DW, Price DL, Hoffman PN (1998) Axonal transport of mutant superoxide dismutase 1 and focal axonal abnormalities in the proximal axons of transgenic mice. *Neurobiol Dis* 5:27-35.

Bray D (1997) The riddle of slow axonal transport - an introduction. *Trends Cell Biol* 7:379.

Brown A (1998) Contiguous phosphorylated and non-phosphorylated domains along axonal neurofilaments. *J Cell Sci* 111 (Pt 4):455-467.

Brown A (2000) Slow axonal transport: stop and go traffic in the axon. *Nat Rev Mol Cell Biol* 1:153-156.

Brownlees J, Ackerley S, Grierson AJ, Jacobsen NJ, Shea K, Anderton BH, Leigh PN, Shaw CE, Miller CC (2002) Charcot-Marie-Tooth disease neurofilament mutations disrupt neurofilament assembly and axonal transport. *Hum Mol Genet* 11:2837-2844.

Bruijn LI, Cudkowicz M (2006) Therapeutic targets for amyotrophic lateral sclerosis: current treatments and prospects for more effective therapies. *Expert Rev Neurother* 6:417-428.

Bruijn LI, Miller TM, Cleveland DW (2004) Unraveling the mechanisms involved in motor neuron degeneration in ALS. *Annu Rev Neurosci* 27:723-749.

Burton PR, Wentz MA (1992) Neurofilaments are prominent in bullfrog olfactory axons but are rarely seen in those of the tiger salamander, Ambystoma tigrinum. *J Comp Neurol* 317:396-406.

Cairns NJ, Perry RH, Jaros E, Burn D, McKeith IG, Lowe JS, Holton J, Rossor MN, Skullerud K, Duyckaerts C, Cruz-Sanchez FF, Lantos PL (2003) Patients with a novel neurofilamentopathy: dementia with neurofilament inclusions. *Neurosci Lett* 341:177-180.

Cairns NJ, Zhukareva V, Uryu K, Zhang B, Bigio E, Mackenzie IR, Gearing M, Duyckaerts C, Yokoo H, Nakazato Y, Jaros E, Perry RH, Lee VM, Trojanowski JQ (2004) alpha-internexin is present in the pathological inclusions of neuronal intermediate filament inclusion disease. *Am J Pathol* 164:2153-2161.

Carden MJ, Schlaepfer WW, Lee VM (1985) The structure, biochemical properties, and immunogenicity of neurofilament peripheral regions are determined by phosphorylation state. *J Biol Chem* 260:9805-9817.

Carden MJ, Trojanowski JQ, Schlaepfer WW, Lee VM (1987) Two-stage expression of neurofilament polypeptides during rat neurogenesis with early establishment of adult phosphorylation patterns. *J Neurosci* 7:3489-3504.

Chan WK, Dickerson A, Ortiz D, Pimenta AF, Moran CM, Motil J, Snyder SJ, Malik K, Pant HC, Shea TB (2004) Mitogen-activated protein kinase regulates neurofilament axonal transport. *J Cell Sci* 117:4629-4642.

Collard JF, Cote F, Julien JP (1995) Defective axonal transport in a transgenic mouse model of amyotrophic lateral sclerosis. *Nature* 375:61-64.

Coulombe PA, Tong X, Mazzalupo S, Wang Z, Wong P (2004) Great promises yet to be fulfilled: defining keratin intermediate filament function in vivo. *Eur J Cell Biol* 83:735-746.

De Jonghe P, Mersivanova I, Nelis E, Del Favero J, Martin JJ, Van Broeckhoven C, Evgrafov O, Timmerman V (2001) Further evidence that neurofilament light chain gene mutations can cause Charcot-Marie-Tooth disease type 2E. *Ann Neurol* 49:245-249.

Ding J, Liu JJ, Kowal AS, Nardine T, Bhattacharya P, Lee A, Yang Y (2002) Microtubule-associated protein 1B: a neuronal binding partner for gigaxonin. *J Cell Biol* 158:427-433.

Drager UC, Hofbauer A (1984) Antibodies to heavy neurofilament subunit detect a subpopulation of damaged ganglion cells in retina. *Nature* 309:624-626.

Elder GA, Friedrich VL, Jr., Margita A, Lazzarini RA (1999) Age-related atrophy of motor axons in mice deficient in the mid-sized neurofilament subunit. *J Cell Biol* 146:181-192.

Elder GA, Friedrich VL, Jr., Bosco P, Kang C, Gourov A, Tu PH, Lee VM, Lazzarini RA (1998a) Absence of the mid-sized neurofilament subunit decreases axonal calibers, levels of light neurofilament (NF-L), and neurofilament content. *J Cell Biol* 141:727-739.

Elder GA, Friedrich VL, Jr., Kang C, Bosco P, Gourov A, Tu PH, Zhang B, Lee VM, Lazzarini RA (1998b) Requirement of heavy neurofilament subunit in the development of axons with large calibers. *J Cell Biol* 143:195-205.

Evgrafov OV, Mersiyanova I, Irobi J, Van Den Bosch L, Dierick I, Leung CL, Schagina O, Verpoorten N, Van Impe K, Fedotov V, Dadali E, Auer-Grumbach M, Windpassinger C, Wagner K, Mitrovic Z, Hilton-Jones D, Talbot K, Martin JJ, Vasserman N, Tverskaya S, Polyakov A, Liem RK, Gettemans J, Robberecht W, De Jonghe P, Timmerman V (2004) Mutant small heat-shock protein 27 causes axonal Charcot-Marie-Tooth disease and distal hereditary motor neuropathy. *Nat Genet* 36:602-606.

Eyer J, Peterson A (1994) Neurofilament-deficient axons and perikaryal aggregates in viable transgenic mice expressing a neurofilament-beta-galactosidase fusion protein. *Neuron* 12:389-405.

Faussone-Pellegrini MS, Matini P, DeFelici M (1999) The cytoskeleton of the myenteric neurons during murine embryonic life. *Anat Embryol (Berl)* 199:459-469.

Fernyhough P, Schmidt RE (2002) Neurofilaments in diabetic neuropathy. *Int Rev Neurobiol* 50:115-144.

Friede RL, Samorajski T (1970) Axon caliber related to neurofilaments and microtubules in sciatic nerve fibers of rats and mice. *Anat Rec* 167:379-387.

Galbraith JA, Reese TS, Schlief ML, Gallant PE (1999) Slow transport of unpolymerized tubulin and polymerized neurofilament in the squid giant axon. *Proc Natl Acad Sci U S A* 96:11589-11594.

Garcia ML, Lobsiger CS, Shah SB, Deerinck TJ, Crum J, Young D, Ward CM, Crawford TO, Gotow T, Uchiyama Y, Ellisman MH, Calcutt NA, Cleveland DW (2003) NF-M is an essential target for the myelin-directed "outside-in" signaling cascade that mediates radial axonal growth. *J Cell Biol* 163:1011-1020.

Geisler N, Vandekerckhove J, Weber K (1987) Location and sequence characterization of the major phosphorylation sites of the high molecular mass neurofilament proteins M and H. *FEBS Lett* 221:403-407.

Georgiou DM, Zidar J, Korosec M, Middleton LT, Kyriakides T, Christodoulou K (2002) A novel NF-L mutation Pro22Ser is associated with CMT2 in a large Slovenian family. *Neurogenetics* 4:93-96.

Giasson BI, Mushynski WE (1997) Developmentally regulated stabilization of neuronal intermediate filaments in rat cerebral cortex. *Neurosci Lett* 229:77-80.

Gibb BJ, Brion JP, Brownlees J, Anderton BH, Miller CC (1998) Neuropathological abnormalities in transgenic mice harbouring a phosphorylation mutant neurofilament transgene. *J Neurochem* 70:492-500.

Gidalevitz T, Ben-Zvi A, Ho KH, Brignull HR, Morimoto RI (2006) Progressive disruption of cellular protein folding in models of polyglutamine diseases. *Science* 311:1471-1474.

Goldstein ME, Sternberger LA, Sternberger NH (1987) Varying degrees of phosphorylation determine microheterogeneity of the heavy neurofilament polypeptide (Nf-H). *J Neuroimmunol* 14:135-148.

Gotow T (2000) Neurofilaments in health and disease. *Med Electron Microsc* 33:173-199.

Grant P, Pant HC (2000) Neurofilament protein synthesis and phosphorylation. *J Neurocytol* 29:843-872.

Gros-Louis F, Lariviere R, Gowing G, Laurent S, Camu W, Bouchard JP, Meininger V, Rouleau GA, Julien JP (2004) A frameshift deletion in peripherin gene associated with amyotrophic lateral sclerosis. *J Biol Chem.*

Hall GF, Chu B, Lee S, Liu Y, Yao J (2000) The single neurofilament subunit of the lamprey forms filaments and regulates axonal caliber and neuronal size in vivo. *Cell Motil Cytoskeleton* 46:166-182.

Helfand BT, Loomis P, Yoon M, Goldman RD (2003) Rapid transport of neural intermediate filament protein. *J Cell Sci* 116:2345-2359.

Hesse M, Magin TM, Weber K (2001) Genes for intermediate filament proteins and the draft sequence of the human genome: novel keratin genes and a surprisingly high number of pseudogenes related to keratin genes 8 and 18. *J Cell Sci* 114:2569-2575.

Hirano A (1991) Cytopathology of amyotrophic lateral sclerosis. In: Advances in Neurology: Amyotrophic lateral sclerosis and other motor neuron diseases (Rowland LP, ed), pp 91-101. New York: Raven Press.

Hirokawa N, Terada S, Funakoshi T, Tekeda S (1997) Slow axonal transport: the subunit transport model. *Trends Cell Biol* 7:384-388.

Hoffman PN, Lasek RJ (1975) The slow component of axonal transport. Identification of major structural polypeptides of the axon and their generality among mammalian neurons. *J Cell Biol* 66:351-366.

Hoffman PN, Griffin JW, Price DL (1984) Control of axonal caliber by neurofilament transport. *J Cell Biol* 99:705-714.

Hoffman PN, Griffin JW, Gold BG, Price DL (1985) Slowing of neurofilament transport and the radial growth of developing nerve fibers. *J Neurosci* 5:2920-2929.

Hoffman PN, Cleveland DW, Griffin JW, Landes PW, Cowan NJ, Price DL (1987) Neurofilament gene expression: a major determinant of axonal caliber. *Proc Natl Acad Sci U S A* 84:3472-3476.

Hursh J (1939) Conduction velocity and diameter of nerve fibers. *Am J Physiol* 127:131-139.

Jacobs AJ, Kamholz J, Selzer ME (1995) The single lamprey neurofilament subunit (NF-180) lacks multiphosphorylation repeats and is expressed selectively in projection neurons. *Brain Res Mol Brain Res* 29:43-52.

Jacomy H, Zhu Q, Couillard-Despres S, Beaulieu JM, Julien JP (1999) Disruption of type IV intermediate filament network in mice lacking the neurofilament medium and heavy subunits. *J Neurochem* 73:972-984.

Jin LQ, Zhang G, Selzer ME (2005) Lamprey neurofilaments contain a previously unreported 50-kDa protein. *J Comp Neurol* 483:403-414.

Jones SM, Williams RC, Jr. (1982) Phosphate content of mammalian neurofilaments. *J Biol Chem* 257:9902-9905.

Jordanova A, De Jonghe P, Boerkoel CF, Takashima H, De Vriendt E, Ceuterick C, Martin JJ, Butler IJ, Mancias P, Papasozomenos S, Terespolsky D, Potocki L, Brown CW, Shy M, Rita DA, Tournev I, Kremensky I, Lupski JR, Timmerman V (2003) Mutations in the neurofilament light chain gene (NEFL) cause early onset severe Charcot-Marie-Tooth disease. *Brain* 126:590-597.

Josephs KA, Holton JL, Rossor MN, Braendgaard H, Ozawa T, Fox NC, Petersen RC, Pearl GS, Ganguly M, Rosa P, Laursen H, Parisi JE, Waldemar G, Quinn NP, Dickson DW, Revesz T (2003) Neurofilament inclusion body disease: a new proteinopathy? *Brain* 126:2291-2303.

Julien JP (1999) Neurofilament functions in health and disease. *Curr Opin Neurobiol* 9:554-560.

Julien JP, Mushynski WE (1982) Multiple phosphorylation sites in mammalian neurofilament polypeptides. *J Biol Chem* 257:10467-10470.

Julien JP, Mushynski WE (1983) The distribution of phosphorylation sites among identified proteolytic fragments of mammalian neurofilaments. *J Biol Chem* 258:4019-4025.

Julien JP, Mushynski WE (1998) Neurofilaments in health and disease. *Prog Nucleic Acid Res Mol Biol* 61:1-23.

Jung C, Yabe JT, Shea TB (2000a) C-terminal phosphorylation of the high molecular weight neurofilament subunit correlates with decreased neurofilament axonal transport velocity. *Brain Res* 856:12-19.

Jung C, Yabe JT, Lee S, Shea TB (2000b) Hypophosphorylated neurofilament subunits undergo axonal transport more rapidly than more extensively phosphorylated subunits in situ. *Cell Motil Cytoskeleton* 47:120-129.

Kim OJ, Ariano MA, Lazzarini RA, Levine MS, Sibley DR (2002) Neurofilament-M interacts with the D1 dopamine receptor to regulate cell surface expression and desensitization. *J Neurosci* 22:5920-5930.

Lariviere RC, Julien JP (2004) Functions of intermediate filaments in neuronal development and disease. *J Neurobiol* 58:131-148.

Lasek RJ, Phillips L, Katz MJ, Autilio-Gambetti L (1985) Function and evolution of neurofilament proteins. *Ann N Y Acad Sci* 455:462-478.

Lavedan C, Buchholtz S, Nussbaum RL, Albin RL, Polymeropoulos MH (2002) A mutation in the human neurofilament M gene in Parkinson's disease that suggests a role for the cytoskeleton in neuronal degeneration. *Neurosci Lett* 322:57-61.

Lee MK, Cleveland DW (1996) Neuronal intermediate filaments. *Annu Rev Neurosci* 19:187-217.

Lee MK, Xu Z, Wong PC, Cleveland DW (1993) Neurofilaments are obligate heteropolymers in vivo. *J Cell Biol* 122:1337-1350.

Lee VM, Carden MJ, Schlaepfer WW, Trojanowski JQ (1987) Monoclonal antibodies distinguish several differentially phosphorylated states of the two largest rat neurofilament subunits (NF-H and NF-M) and demonstrate their existence in the normal nervous system of adult rats. *J Neurosci* 7:3474-3488.

Lee VM, Otvos L, Jr., Carden MJ, Hollosi M, Dietzschold B, Lazzarini RA (1988) Identification of the major multiphosphorylation site in mammalian neurofilaments. *Proc Natl Acad Sci U S A* 85:1998-2002.

Leung CL, He CZ, Kaufmann P, Chin SS, Naini A, Liem RK, Mitsumoto H, Hays AR (2004) A pathogenic peripherin gene mutation in a patient with amyotrophic lateral sclerosis. *Brain Pathol* 14:290-296.

Lewis SE, Nixon RA (1988) Multiple phosphorylated variants of the high molecular mass subunit of neurofilaments in axons of retinal cell neurons: characterization and evidence for their differential association with stationary and moving neurofilaments. *J Cell Biol* 107:2689-2701.

Liem RK, Yen SH, Salomon GD, Shelanski ML (1978) Intermediate filaments in nervous tissues. *J Cell Biol* 79:637-645.

Liu Q, Xie F, Siedlak SL, Nunomura A, Honda K, Moreira PI, Zhua X, Smith MA, Perry G (2004) Neurofilament proteins in neurodegenerative diseases. *Cell Mol Life Sci* 61:3057-3075.

Marszalek JR, Williamson TL, Lee MK, Xu Z, Hoffman PN, Becher MW, Crawford TO, Cleveland DW (1996) Neurofilament subunit NF-H modulates axonal diameter by selectively slowing neurofilament transport. *J Cell Biol* 135:711-724.

McQuarrie IG, Brady ST, Lasek RJ (1989) Retardation in the slow axonal transport of cytoskeletal elements during maturation and aging. *Neurobiol Aging* 10:359-365.

Menzies FM, Grierson AJ, Cookson MR, Heath PR, Tomkins J, Figlewicz DA, Ince PG, Shaw PJ (2002) Selective loss of neurofilament expression in Cu/Zn superoxide dismutase (SOD1) linked amyotrophic lateral sclerosis. *J Neurochem* 82:1118-1128.

Mersiyanova IV, Perepelov AV, Polyakov AV, Sitnikov VF, Dadali EL, Oparin RB, Petrin AN, Evgrafov OV (2000) A new variant of Charcot-Marie-Tooth disease type 2 is probably the result of a mutation in the neurofilament-light gene. *Am J Hum Genet* 67:37-46.

Mills RG, Minamide LS, Yuan A, Bamburg JR, Bray JJ (1996) Slow axonal transport of soluble actin with actin depolymerizing factor, cofilin, and profilin suggests actin moves in an unassembled form. *J Neurochem* 67:1225-1234.

Momeni P, Cairns NJ, Perry RH, Bigio E, Gearing M, Singleton AB, Hardy J (2005) Mutation analysis of patients with neuronal intermediate filament inclusion disease (NIFID). *Neurobiol Aging.*

Monaco S, Autilio-Gambetti L, Zabel D, Gambetti P (1985) Giant axonal neuropathy: acceleration of neurofilament transport in optic axons. *Proc Natl Acad Sci U S A* 82:920-924.

Monaco S, Wongmongkolrit T, Shearson CM, Patton A, Schaetzle B, Autilio-Gambetti L, Gambetti P, Sayre LM (1990) Giant axonopathy characterized by intermediate location of axonal enlargements and acceleration of neurofilament transport. *Brain Res* 519:73-81.

Mori H, Komiya Y, Kurokawa M (1979) Slowly migrating axonal polypeptides. Inequalities in their rate and amount of transport between two branches of bifurcating axons. *J Cell Biol* 82:174-184.

Nixon RA (1998) Dynamic behavior and organization of cytoskeletal proteins in neurons: reconciling old and new findings. *Bioessays* 20:798-807.

Nixon RA, Logvinenko KB (1986) Multiple fates of newly synthesized neurofilament proteins: evidence for a stationary neurofilament network distributed nonuniformly along axons of retinal ganglion cell neurons. *J Cell Biol* 102:647-659.

Nixon RA, Brown BA, Marotta CA (1982) Posttranslational modification of a neurofilament protein during axoplasmic transport: implications for regional specialization of CNS axons. *J Cell Biol* 94:150-158.

Nixon RA, Lewis SE, Marotta CA (1987) Posttranslational modification of neurofilament proteins by phosphate during axoplasmic transport in retinal ganglion cell neurons. *J Neurosci* 7:1145-1158.

Nixon RA, Paskevich PA, Sihag RK, Thayer CY (1994) Phosphorylation on carboxyl terminus domains of neurofilament proteins in retinal ganglion cell neurons in vivo: influences on regional neurofilament accumulation, interneurofilament spacing, and axon caliber. *J Cell Biol* 126:1031-1046.

Ohara O, Gahara Y, Miyake T, Teraoka H, Kitamura T (1993) Neurofilament deficiency in quail caused by nonsense mutation in neurofilament-L gene. *J Cell Biol* 121:387-395.

Pachter JS, Liem RK (1984) The differential appearance of neurofilament triplet polypeptides in the developing rat optic nerve. *Dev Biol* 103:200-210.

Perez-Olle R, Leung CL, Liem RK (2002) Effects of Charcot-Marie-Tooth-linked mutations of the neurofilament light subunit on intermediate filament formation. *J Cell Sci* 115:4937-4946.

Perez-Olle R, Lopez-Toledano MA, Goryunov D, Cabrera-Poch N, Stefanis L, Brown K, Liem RK (2005) Mutations in the neurofilament light gene linked to Charcot-Marie-Tooth disease cause defects in transport. *J Neurochem* 93:861-874.

Perrone Capano C, Pernas-Alonso R, di Porzio U (2001) Neurofilament homeostasis and motoneurone degeneration. *Bioessays* 23:24-33.

Petzold A (2005) Neurofilament phosphoforms: Surrogate markers for axonal injury, degeneration and loss. *J Neurol Sci* 233:183-198.

Prahlad V, Helfand BT, Langford GM, Vale RD, Goldman RD (2000) Fast transport of neurofilament protein along microtubules in squid axoplasm. *J Cell Sci* 113:3939-3946.

Rao MV, Nixon RA (2002) Neurofilaments. In: Wiley Encyclopedia of molecular medicine, pp 2237-2242: John Wiley and Sons, Inc.

Rao MV, Nixon RA (2003) Defective neurofilament transport in mouse models of amyotrophic lateral sclerosis: a review. *Neurochem Res* 28:1041-1047.

Rao MV, Houseweart MK, Williamson TL, Crawford TO, Folmer J, Cleveland DW (1998) Neurofilament-dependent radial growth of motor axons and axonal organization of neurofilaments does not require the neurofilament heavy subunit (NF-H) or its phosphorylation. *J Cell Biol* 143:171-181.

Rao MV, Yuan A, Campbell J, Kumar A, Veeranna., Nixon RA (2004) The phosphorylated tail domains of neurofilament heavy and medium subunits together regulate neurofilament content but not the rate of slow axonal transport of neurofilaments in optic axons. In: *Society for Neuroscience Annual Meeting* 2004. Washington, DC.

Rao MV, Campbell J, Yuan A, Kumar A, Gotow T, Uchiyama Y, Nixon RA (2003) The neurofilament middle molecular mass subunit carboxyl-terminal tail domains is essential for the radial growth and cytoskeletal architecture of axons but not for regulating neurofilament transport rate. *J Cell Biol* 163:1021-1031.

Rao MV, Garcia ML, Miyazaki Y, Gotow T, Yuan A, Mattina S, Ward CM, Calcutt NA, Uchiyama Y, Nixon RA, Cleveland DW (2002) Gene replacement in mice reveals that the heavily phosphorylated tail of neurofilament heavy subunit does not affect axonal caliber or the transit of cargoes in slow axonal transport. *J Cell Biol* 158:681-693.

Rosen DR, Siddique T, Patterson D, Figlewicz DA, Sapp P, Hentati A, Donaldson D, Goto J, O'Regan JP, Deng HX, et al. (1993) Mutations in Cu/Zn superoxide dismutase gene are associated with familial amyotrophic lateral sclerosis. *Nature* 362:59-62.

Roy S, Coffee P, Smith G, Liem RK, Brady ST, Black MM (2000) Neurofilaments are transported rapidly but intermittently in axons: implications for slow axonal transport. *J Neurosci* 20:6849-6861.

Ruiz-Ederra J, Vecino E (2001) [Neurofilaments in neurodegenerative diseases]. *Arch Soc Esp Oftalmol* 76:699-710.

Sakaguchi T, Okada M, Kitamura T, Kawasaki K (1993) Reduced diameter and conduction velocity of myelinated fibers in the sciatic nerve of a neurofilament-deficient mutant quail. *Neurosci Lett* 153:65-68.

Sanchez I, Hassinger L, Sihag RK, Cleveland DW, Mohan P, Nixon RA (2000) Local control of neurofilament accumulation during radial growth of myelinating axons in vivo. Selective role of site-specific phosphorylation. *J Cell Biol* 151:1013-1024.

Sasaki T, Gotow T, Shiozaki M, Sakaue F, Saito T, Julien JP, Uchiyama Y, Hisanaga S (2006) Aggregate formation and phosphorylation of neurofilament-L Pro22 Charcot-Marie-Tooth disease mutants. *Hum Mol Genet* 15:943-952.

Schmitt FO (1968) Fibrous proteins--neuronal organelles. *Proc Natl Acad Sci U S A* 60:1092-1101.

Shah JV, Cleveland DW (2002) Slow axonal transport: fast motors in the slow lane. *Curr Opin Cell Biol* 14:58-62.

Shaw G, Weber K (1982) Differential expression of neurofilament triplet proteins in brain development. *Nature* 298:277-279.

Sidenius P, Nagel P, Larsen JR, Boye N, Laurberg P (1987) Axonal transport of slow component a in sciatic nerves of hypo- and hyperthyroid rats. *J Neurochem* 49:1790-1795.

Sihag RK, Nixon RA (1990) Phosphorylation of the amino-terminal head domain of the middle molecular mass 145-kDa subunit of neurofilaments. Evidence for regulation by second messenger-dependent protein kinases. *J Biol Chem* 265:4166-4171.

Stein SA, Kirkpatrick LL, Shanklin DR, Adams PM, Brady ST (1991) Hypothyroidism reduces the rate of slow component A (SCa) axonal transport and the amount of transported tubulin in the hyt/hyt mouse optic nerve. *J Neurosci Res* 28:121-133.

Sternberger LA, Sternberger NH (1983) Monoclonal antibodies distinguish phosphorylated and nonphosphorylated forms of neurofilaments in situ. *Proc Natl Acad Sci U S A* 80:6126-6130.

Strong MJ, Strong WL, Jaffe H, Traggert B, Sopper MM, Pant HC (2001) Phosphorylation state of the native high-molecular-weight neurofilament subunit protein from cervical spinal cord in sporadic amyotrophic lateral sclerosis. *J Neurochem* 76:1315-1325.

Szaro BG, Lee VM, Gainer H (1989) Spatial and temporal expression of phosphorylated and non-phosphorylated forms of neurofilament proteins in the developing nervous system of Xenopus laevis. *Brain Res Dev Brain Res* 48:87-103.

Takeda S, Okabe S, Funakoshi T, Hirokawa N (1994) Differential dynamics of neurofilament-H protein and neurofilament-L protein in neurons. *J Cell Biol* 127:173-185.

Terada S, Hirokawa N (2000) Moving on to the cargo problem of microtubule-dependent motors in neurons. *Curr Opin Neurobiol* 10:566-573.

Terada S, Nakata T, Peterson AC, Hirokawa N (1996) Visualization of slow axonal transport in vivo. *Science* 273:784-788.

Theiss C, Napirei M, Meller K (2005) Impairment of anterograde and retrograde neurofilament transport after anti-kinesin and anti-dynein antibody microinjection in chicken dorsal root ganglia. *Eur J Cell Biol* 84:29-43.

Toyoshima I, Kato K, Sugawara M, Wada C, Okawa S, Kobayashi M, Masamune O, Watanabe S (2000) Massive accumulation of M and H subunits of neurofilament proteins in spinal motor neurons of neurofilament deficient Japanese quail, Quv. *Neurosci Lett* 287:175-178.

Trojanowski JQ, Schmidt ML, Shin RW, Bramblett GT, Rao D, Lee VM (1993) Altered tau and neurofilament proteins in neuro-degenerative diseases: diagnostic implications for Alzheimer's disease and Lewy body dementias. *Brain Pathol* 3:45-54.

Uchida A, Brown A (2004) Arrival, reversal, and departure of neurofilaments at the tips of growing axons. *Mol Biol Cell* 15:4215-4225.

Undamatla J, Szaro BG (2001) Differential expression and localization of neuronal intermediate filament proteins within newly developing neurites in dissociated cultures of Xenopus laevis embryonic spinal cord. *Cell Motil Cytoskeleton* 49:16-32.

Wagner OI, Ascano J, Tokito M, Leterrier JF, Janmey PA, Holzbaur EL (2004) The interaction of neurofilaments with the microtubule motor cytoplasmic dynein. *Mol Biol Cell* 15:5092-5100.

Wang L, Brown A (2001) Rapid intermittent movement of axonal neurofilaments observed by fluorescence photobleaching. *Mol Biol Cell* 12:3257-3267.

Wang L, Ho CL, Sun D, Liem RK, Brown A (2000) Rapid movement of axonal neurofilaments interrupted by prolonged pauses. *Nat Cell Biol* 2:137-141.

Waxman SG (1980) Determinants of conduction velocity in myelinated nerve fibers. *Muscle Nerve* 3:141-150.

Willard M, Simon C (1981) Antibody decoration of neurofilaments. J Cell Biol 89:198-205.

Willard M, Simon C (1983) Modulations of neurofilament axonal transport during the development of rabbit retinal ganglion cells. *Cell* 35:551-559.

Williamson TL, Cleveland DW (1999) Slowing of axonal transport is a very early event in the toxicity of ALS-linked SOD1 mutants to motor neurons. *Nat Neurosci* 2:50-56.

Yabe JT, Pimenta A, Shea TB (1999) Kinesin-mediated transport of neurofilament protein oligomers in growing axons. *J Cell Sci* 112:3799-3814.

Yabe JT, Chylinski T, Wang FS, Pimenta A, Kattar SD, Linsley MD, Chan WK, Shea TB (2001) Neurofilaments consist of distinct populations that can be distinguished by C-terminal phosphorylation, bundling, and axonal transport rate in growing axonal neurites. *J Neurosci* 21:2195-2205.

Yamasaki H, Itakura C, Mizutani M (1991) Hereditary hypotrophic axonopathy with neurofilament deficiency in a mutant strain of the Japanese quail. *Acta Neuropathol* (Berl) 82:427-434.

Yamasaki H, Bennett GS, Itakura C, Mizutani M (1992) Defective expression of neurofilament protein subunits in hereditary hypotrophic axonopathy of quail. *Lab Invest* 66:734-743.

Yan Y, Brown A (2005) Neurofilament polymer transport in axons. *J Neurosci* 25:7014-7021.

Yoshihara T, Yamamoto M, Hattori N, Misu K, Mori K, Koike H, Sobue G (2002) Identification of novel sequence variants in the neurofilament-light gene in a Japanese population: analysis of Charcot-Marie-Tooth disease patients and normal individuals. *J Peripher Nerv Syst* 7:221-224.

Yuan A (1997) *Axonal transport of cytoskeletal and associated proteins.* PhD thesis. Dunedin: Otago University.

Yuan A, Nixon RA, Rao MV (2006) Deleting the phosphorylated tail domain of the neurofilament heavy subunit does not alter neurofilament transport rate in vivo. *Neurosci Lett* 393:264-268.

Yuan A, Mills RG, Chia CP, Bray JJ (2000) Tubulin and neurofilament proteins are transported differently in axons of chicken motoneurons. *Cell Mol Neurobiol* 20:623-632.

Yuan A, Rao MV, Kumar A, Julien JP, Nixon RA (2003) Neurofilament transport in vivo minimally requires hetero-oligomer formation. *J Neurosci* 23:9452-9458.

Zackroff RV, Idler WW, Steinert PM, Goldman RD (1982) In vitro reconstitution of intermediate filaments form mammalian neurofilament triplet polypeptides. *Proc Natl Acad Sci U S A* 79:754-757.

Zhang B, Tu P, Abtahian F, Trojanowski JQ, Lee VM (1997) Neurofilaments and orthograde transport are reduced in ventral root axons of transgenic mice that express human SOD1 with a G93A mutation. *J Cell Biol* 139:1307-1315.

Zhu Q, Couillard-Despres S, Julien JP (1997) Delayed maturation of regenerating myelinated axons in mice lacking neurofilaments. *Exp Neurol* 148:299-316.

Zhu Q, Lindenbaum M, Levavasseur F, Jacomy H, Julien JP (1998) Disruption of the NF-H gene increases axonal microtubule content and velocity of neurofilament transport: relief of axonopathy resulting from the toxin beta,beta'-iminodipropionitrile. *J Cell Biol* 143:183-193.

Zuchner S, Vorgerd M, Sindern E, Schroder JM (2004) The novel neurofilament light (NEFL) mutation Glu397Lys is associated with a clinically and morphologically heterogeneous type of Charcot-Marie-Tooth neuropathy. *Neuromuscul Disord* 14:147-157.

In: New Research on Neurofilament Proteins
Editor: Roland K. Arlen, pp. 81-97

ISBN: 1-60021-396-0
© 2007 Nova Science Publishers, Inc.

Chapter IV

Neurofilament Changes in Multiple Sclerosis

*Alastair Wilkins**
Department of Clinical Neurosciences and Centre for Brain Repair,
University of Cambridge, Forvie Site, Robinson Way, Cambridge, UK

Abstract

Multiple sclerosis is a disease of the central nervous system characterised by episodes of neurological dysfunction which often recover, usually followed some years later by progressive and irreversible decline. Lesions of multiple sclerosis are characterised by varying degrees of inflammation, myelin and oligodendrocyte loss, astrogliosis and axonal pathology. Axonal loss is seen in progressive phases of the disease and appears to correlate well with clinical disability. Prior to loss of axons, pathological specimens have revealed changes in the immunohistochemical phenotype of axons. Specifically there may be evidence of dephosphorylation of neurofilaments within axons and transection of axons leading to the formation of axonal spheroids which are rich in dephosphorylated neurofilaments. Evidence of axonal transport defects may also be found in lesions with accumulation of amyloid precursor protein within the axon. Mechanisms of axonal pathology in multiple sclerosis remain unknown, but it is likely that inflammation causes some degree of damage in the acute phases. However there is evidence that axonal loss may continue even in the absence of inflammation. It has been postulated that axonal loss in this situation occurs due to loss of trophic support from myelin and oligodendrocytes. Oligodendrocytes and myelin are known to provide trophic support for axons and specifically can influence phosphorylated neurofilament levels. The precise mechanisms are unknown, but recent evidence suggests a combination of contact mediated and soluble factors may increase neurofilament phosphorylation and promote axonal protection. Knowledge of such mechanisms may lead to improved therapies to prevent progressive disease. This review will discuss axonal changes in multiple sclerosis, specifically alteration in neurofilament phosphorylation states, and potential mechanisms of axonal protection.

Abbreviations

BDNF	brain-derived neurotrophic factor
CNPase	2',3'-cyclic nucleotide 3'-phosphodiesterase
CNS	central nervous system
CSF	cerebrospinal fluid
GDNF	glial cell-line derived neurotrophic factor
IF	intermediate filament
IGF-1	insulin-like growth factor type-1
MAG	myelin associated glycoprotein
MBP	myelin basic protein
MOG	myelin-oligodendrocyte glycoprotein
NF	neurofilament
NGF	nerve growth factor
NO	nitric oxide
PLP	proteolipid protein

Introduction

Multiple sclerosis is a common disease of the central nervous system (CNS) which presents significant challenges to the clinician treating the disease and scientists researching it. Once thought the archetypal inflammatory CNS disorder, evidence now suggests the pathophysiology is complex and possibly occurs via multiple mechanisms including neurodegeneration (Compston, 2006). The complexity of the disease is born out by the inadequacy of current treatments; and the progressive phase of the disease, during which patients accrue disability relentlessly, has no effective therapy.

Axon pathology in multiple sclerosis was first described by Charcot in his classical reports of the disease over 100 years ago, but the focus on inflammatory demyelinating plaques and their pathogenesis led to a paucity of research on the subject until several reports in the late 1990s re-emphasised this aspect of pathology. Most notably, Trapp and colleagues (1998) demonstrated axonal transection and changes in neurofilament phosphorylation states

---

[*] Correspondence: Alastair Wilkins, Department of Clinical Neurosciences and Centre for Brain Repair, University of Cambridge, Forvie Site, Robinson Way, Cambridge CB2 2PY UK Tel: +44 1223 331186, Fax: +44 1223 331174; e-mail: aw255@cam.ac.uk

in axons within acute lesions of multiple sclerosis. Subsequent work has shown further axonal changes and has revealed that axonal loss belies progression of the disease.

This review will discuss the evidence behind axonal pathology in multiple sclerosis, focusing on neurofilament changes; discuss potential mechanisms of the process and possible neuroprotective therapies in the disease.

The Natural History of Multiple Sclerosis

Multiple sclerosis is generally characterised by episodes of acute neurological dysfunction which usually recover, followed at some stage by slow and insidious progression and development of fixed neurological deficits (Compston and Coles, 2002). Most patients experience an initial relapsing and remitting preliminary course of the disease. During such relapses neurological dysfunction is thought to be due to the effect of inflammatory cytokines and conduction block resulting from disruption of myelin (Coles et al. 1999). Myelin is produced by oligodendrocytes in the central nervous system and allows for rapid saltatory conduction to occur via the formation of nodes of Ranvier. At such nodes there is grouping of sodium channels allowing for efficient action potential propagation. Furthermore myelin itself offers high resistance and low capacitance which again improves saltatory conduction. In the absence of myelin, axonal conduction is severely impaired and conduction of electrical impulses between internodes fails by virtue of myelin loss and a redistribution of sodium channels along the length of the axon (Craner et al. 2003; Craner et al. 2004).

Mechanisms for remission from attacks are rather poorly understood, but involve resolution of inflammation, reorganisation of axonal channels and limited remyelination (Smith et al. 1979). Indeed current therapies for the disease aim at reducing inflammation during the acute phase of the disease with drugs such as β-interferon (Jacobs et al. 2000). However most patients, in due course, will enter a phase of the disease in which remission does not occur and the disease progresses inexorably. During this progressive phase of the disease disability correlates with radiological markers of axon loss, including magnetic resonance spectroscopic analysis of NAA and brain atrophy (Davie et al. 1997). Reasons as to why remission does not occur at this stage of the disease are the subject of extensive research, but a long-term failure of remyelination and lack of a supportive environment for the axon seem likely. Failure of effective remyelination is thought to be due to depletion of remyelinating cells, quiescence of oligodendrocyte precursor cells and axonal inhibitory signals (Lubetzki et al. 2005).

At the present time once the progressive phase of the disease has been entered, there are no effective therapies for the disease, which would appear to reflect the inability of therapeutic agents to prevent axonal degeneration. Indeed a striking failure of current treatments for multiple sclerosis has been there poor efficacy in preventing accumulation of disability in later stages of the disease (Kappos, 2004). The main reason for this appears to relate to the mode of action of drugs such as β-interferon which target the inflammatory response. Epidemiological data concerning multiple sclerosis has, however, shown that inflammation is unlikely to be the sole cause for clinical deterioration in the progressive phases of the disease.

Specifically the European database for multiple sclerosis (EDMUS) database has been highly informative. This has revealed that once a certain level of disability has been reached in patient cohorts, further disease progression occurs at the same rate in those who do not have superimposed relapses, compared to those who do (Confavreux et al. 2000). This is in contrast to the clinical course early in the disease process, which does seem to be dependent on relapse rate in determining early disease rate. More recent evidence has suggested that there may be dissociation of the inflammatory and degenerative processes even from disease onset, as evidenced by similarities in rates of progression of those patients with primary progressive disease, in which the disease does not commence with a period of relapses and remissions, and those with secondary progressive disease (Confavreux and Vukusic, 2006). Furthermore evidence from treatment trials concurs with the concept that disease progression may be partially independent of inflammatory activity. CAMPATH-1H, a humanised monoclonal antibody, is highly effective in abolishing inflammatory relapses in patients with multiple sclerosis, but has no effect on disease progression in those patients with existing disability (Coles et al. 1999).

From these studies has emerged, therefore, the notion of two processes occurring during the pathogenesis of multiple sclerosis. The idea that the inflammatory and the neurodegenerative processes may run in parallel and may be independent of each other challenges existing concepts about the disease and the way in which it should be treated. The search for therapies which tackle the neurodegenerative component of disease progression is an important concept. At the heart of this is the need to understand axonal pathology in the disease and mechanisms underlying axon protection.

Axonal Pathology in MS

The classic histological features of acute lesions of multiple sclerosis include perivascular infiltrates of small lymphocytes, loss of myelin and oligodendrocytes, myelin-laden macrophages and axonal damage (Frohman et al. 2006). However there may be extensive variability between lesions. Some authors have recognised 4 distinct types of acute lesions based on pathological studies (Lucchinetti et al. 1996; Lucchinetti et al. 2000). These types are felt to reflect potential heterogeneity in disease pathogenesis. Patterns I and II show similarities to T-cell-mediated or T-cell plus antibody-mediated autoimmune encephalomyelitis. Type I is characterised by demyelination and the presence of macrophages and cytokines, such as tumour necrosis factor-α (TNFα). Type II is similar but shows immunoglobulin and complement activity. Type III, in contrast, lacks immunoglobulin deposition, but shows oligodendrocyte dysfunction and demyelination, whereas type IV is characterised by oligodendrocyte apoptosis through DNA fragmentation. Thus types III and IV are suggestive of a primary oligodendrocyte dystrophy, reminiscent of virus- or toxin-induced demyelination rather than autoimmunity. The reasons for these differences and the importance in disease phenotypes have yet to be determined. However, a further demyelinating disease of the central nervous system, Devic's disease or neuromyelitis optica, which is a distinct variant from multiple sclerosis shows marked antibody mediated tissue injury, is characterised by the presence of a specific circulating antibody in the serum and typically responds to plasma exchange therapy (Scolding, 2005). This pathological and phenotypic subtype of CNS demyelination with a good response to a certain therapy distinguishes it from other inflammatory disease and raises the future possibility of tailoring therapies to immunopathological subtypes of disease.

In chronic active lesions of MS the histological appearances may be somewhat different with hypertophic astocytes and degenerating axons, as well as perivascular cuffs of infiltrating cells and lipid-laden macrophages. These lesions may show some degree of remyelination and recruitment of cells of the oligodendrocyte lineage, although remyelination is generally poor and incomplete (Dubois-Dalq et al. 2005). This failure of repair in chronic lesions is thought to be a major reason for disease progression and much active research is targeted at attempts to enhance remyelination (Zhao et al, 2005). Chronic inactive lesions of MS are characterised by astrogliosis, a paucity of oligodendrocytes and reduced numbers of demyelinated axons.

Axonal pathology was described by Charcot in his initial histopathological descriptions of the disease at the end of the nineteenth century, but it was not until over one hundred years later that the subject of axonopathy in MS became the focus of active research. Histopathological analyses of chronic multiple sclerosis lesions have revealed significant reductions in axon density (Lovas et al. 2000; DeLuca et al. 2004). In addition, a number of axonal markers have now been used which highlight dysfunction of the axon prior to degeneration. The two markers in commonest use are: amyloid precursor protein accumulation (APP) and the presence of dephosphorylated neurofilaments.

APP is rapidly transported down axons and normal axons do not stain for the protein. The presence of APP staining in axon tracts implies disruption of fast axonal transport and thus dysfunction of the axon. APP accumulation occurs in both acute active lesions and in chronic lesions of MS (Ferguson et al. 1997; Bitsch et al. 2000; Kuhlmann et al. 2002). APP accumulation may also occur in 'normal appearing white matter' of multiple sclerosis tissue, which has again added to evidence that axonopathy maybe widespread and not exclusively related to inflammation (Kornek et al. 2000). However, APP accumulation within axons is most marked within active inflammatory lesions and appears to correlate with numbers of infiltrating inflammatory cells (Bitsch et al. 2000). It may be, therefore, that APP accumulation within axons is marking acute disruption of axonal function and may be the earliest sign of axonal dysfunction within a lesion. It is unknown the extent to which the process of APP accumulation within axons is reversible, but this concept is likely to be of vital importance when considering potential therapies for axonopathy in the disease.

Neurofilament dephosphorylation within axons has been well documented in multiple sclerosis and will be discussed in detail below.

Other axonal changes have been reported, notably changes in ion channel distribution. In normal myelinated axons sodium channels are clustered at high density at nodes of Ranvier, facilitating effective saltatory conduction. Recent work on sodium channel distribution in MS tissue has revealed changes in channel distributions associated with areas of axonal injury (Craner et al. 2004). In axons positive for APP staining, corresponding to axonal dysfunction, the sodium channel subtype ($Na_v1.6$) which usually clusters at the node of Ranvier is found rather diffusely along the axon. Furthermore the channel co-localises with the sodium/calcium exchanger. Increased and persistent sodium influx through $Na_v1.6$ channels may then lead to reversal of the sodium/channel exchange mechanism and an accumulation of intraxonal calcium, leading to axonal injury (Waxman *et al.* 2004). Indeed similar mechanisms have been postulated in experimental models of inflammatory central nervous system disease. Rat dorsal roots exposed to nitric oxide degenerate when electrically activated, a process which is attenuated by sodium channel blockers, such as flecainide (Smith *et al.* 2000; Kapoor *et al.* 2003).

Axonal dysfunction can also be studied by other techniques. Great advances have been made in recent years in imaging the brain and spinal cord. This has led to greater diagnostic clarity but has also provided insight into pathophysiological processes that may be occurring throughout the duration of the disease. Specifically, magnetic resonance spectroscopy (MRS) is a technique which allows quantification of certain tissue metabolites in the brains of live patients. N-acetyl aspartate (NAA) is found predominantly in neurons and the spectroscopic detection of the compound is a marker of axonal density in the nervous system (Bjartmar *et al.* 2000). Reductions in brain NAA levels have been detected in acute active lesions of MS, chronic active lesions and so-called normal appearing white matter (NAWM) of patients with the disease (Davie *et al.* 1995; Leary *et al.* 1999). Furthermore persistent neurological disability correlates to reductions in NAA levels, an observation which has led to the hypothesis that permanent disability in MS is caused by axonal loss. An '*in vivo*' marker of axonal integrity is a vital component of future trials which aim to address neurodegenerative processes in the disease.

Indeed, the search for 'biomarkers' in multiple sclerosis has been intense in an attempt to correlate potential therapies with effects on pathology in patients. One such marker is the presence of neurofilaments with the cerebrospinal fluid (CSF; Petzold, 2005). It has been postulated that the presence of these molecules within the CSF may be an indicator of axon destruction within the CNS: when axons degenerate neurofilaments are released into the extracellular fluid space and from there enter the CSF. The precise relationship between CSF concentrations of neurofilaments and the degree of axonal degeneration is unknown. Although more work is required on changes in neurofilament levels in the CSF as the disease progresses, there is a suggestion that levels increase during progressive phases of multiple sclerosis (Petzold *et al.* 2005).

Neurofilaments and MS

As discussed, changes in neurofilament phosphorylation states have been detected in axons and taken as evidence of pathological dysfunction. Below I will discuss axonal structure, changes in neurofilament phosphorylation levels in multiple sclerosis and the influence of myelin and oligodendrocytes on neurofilament phosphorylation in relation to putative changes occurring in multiple sclerosis.

The axonal cytoskeleton is composed of scaffolding filaments: actin microfilaments, microtubules and intermediate filaments (IFs). Neurofilaments (NFs) are neuron-specific IFs which consist of three subunits – a light chain (NF-L), a medium chain (NF-M) and a heavy chain (NF-H) differing in molecular weight by virtue of their hypervariable tail domains (Fuchs and Weber, 1994). The C-terminal region of NF-M and NF-H contain 10-15 and 40-

50 lysine-serine-proline (KSP) repeats respectively acting as potential phosphorylation sites (Veeranna *et al.* 1998). NF-H is the most extensively phosphorylated protein in the human brain and the process of phosphorylation and dephosphorylation is highly complex. In brief, phosphorylation occurs via a number of kinases and is predominantly regulated within the axonal compartment (Grant and Pant, 2000). During development, NF phosphorylation appears to be a key factor in triggering NF accumulation and thus the formation of axons (Sanchez *et al.* 2000). Phosphorylation of NFs also leads to slowing of NF transport, promotion of NF alignment and an increase in inter-NF spacing, all of which contribute to axon structure, function and stability (Leterrier *et al.* 1996; Yabe *et al.* 2001). Virtually all NF-H KSP sites in mature, healthy axons are phosphorylated *in vivo*, a process which occurs exclusively in axons (Elhanany *et al.* 1994).

Neurofilaments are slowly transported down the axon and the velocity of transport is inversely proportional to degree of phosphorylation (Nixon *et al.* 1994). It is thought that when neurofilaments reach the axon terminus they are then degraded through protease dependent mechanisms. High levels of phosphorylation within neurofilaments are however protective against proteolysis (Goldstein *et al.* 1987). The corollary to which suggests that neurofilaments that have become dephosphorylated may be more susceptible to proteolytic damage and thus degeneration of the axon. Whether this process underlies axon degeneration in multiple sclerosis is unknown.

Neurofilament dephosphorylation has been used as a marker for axonal pathology in multiple sclerosis and in experimental models of demyelination. The precise relationship between neurofilament dephosphorylation and axonal degeneration is not known. Nor is it known exactly why neurofilaments become dephosphorylated.

Antibodies directed against different phosphoisoforms of neurofilament have been used to study the process in multiple sclerosis. Most commonly, antibodies against dephosphorylated neurofilaments (SMI32; Sternberger Monoclonals, USA) have been used and shown to mark axonal abnormalities in the disease (Trapp *et al.* 1998). In this study, neurofilament dephosphorylation was seen in both active inflammatory lesions and also in chronic active lesions. SMI32 also marked terminal axonal swelling or 'spheroids' which represent transected axons and thus irreversible axonal destruction. Interestingly, quantification of terminal axonal spheroids identified a strong correlation between axonal transection and the inflammatory activity of the lesion. This provides evidence that axonopathy in the disease may occur early and be related, to some extent, to inflammation.

Mechanisms of inflammatory axon destruction are likely to be complex, but a role for nitric oxide (NO) has emerged from experimental studies. NO serves many physiological functions within the nervous system, including roles in synaptic plasticity, long-term potentiation and neurotransmitter release, but at sites of inflammation high concentrations of NO are thought to mediate cell death (Smith and Lassmann, 2002). Microglial derived NO has been shown to be neurotoxic *in vitro* (Bal-Price and Brown, 2001; Golde *et al.* 2001). Mechanisms underlying this neurotoxicity are complex, but NO activates a number of intracellular signaling pathways including MAPkinases and may ultimately lead to neuronal death via inhibition of mitochondrial respiration (Brown and Borutaite, 2002; Ghatan *et al.* 2000). Recent *in vivo* work has shown that NO may also mediate activity dependent axon destruction (Smith *et al.* 2001). Furthermore pathological studies have demonstrated activity of inducible nitric oxide sythetase (iNOS) within acute inflammatory lesions of multiple sclerosis (Bagasra et al. 1995).

NO has also been shown to have effects on neurofilament phosphorylation levels within axons in culture (Wilkins and Compston, 2005). In this system, activation of p38 MAPkinase by NO leads to axonal destruction and reduced levels of phosphorylated neurofilaments (via inhibition of MAPkinase/Erk signaling). Activation of MAPkinase/Erk and inhibition of p38 MAPkinase signaling by growth factors lead to attenuation of NO mediated axonal destruction.

Evidence for chronic axonal attrition has accumulated in recent years with pathological studies demonstrating reduction in axonal densities in chronic inactive plaques (Lovas et al. 2000; DeLuca et al. 2004). It is unclear whether axon drop-out occurs in these late stages as a result of previous inflammatory damage to axons; as a result of low grade inflammation occurring causing damage to already vulnerable demyelinated axons; as a result of loss of trophic environment for axons to survive; or as part of a completely independent neurodegenerative process. Indeed recent evidence has shown that there may be poor correlation between plaque load and axon loss in post-mortem specimens of multiple sclerosis tissue (deLuca et al. 2006). Certain axonal tracts also appear more susceptible to injury in multiple sclerosis (deLuca et al. 2004). These studies, although only providing a 'snap shot' of a dynamic disease, argue that demyelination is not the primary determinant of axon loss and axonal degeneration may be driven by non-inflammatory mechanisms.

How to tie up these seemingly disparate pathological mechanisms? One might hypothesise that inflammation, causing widespread tissue damage, contributes to early axon destruction via molecules such as nitric oxide. However, destruction of oligodendrocytes and myelin also occurs at this stage and may 'set up' the later process of chronic axonal drop-out. Myelin mutant studies have shown that myelin is an important determinant of long term axonal survival (see below). Thus myelin and oligodendrocyte loss may leave axons particularly susceptible to degeneration via this mechanism. Furthermore such axons may be rendered particularly susceptible to very low levels of inflammation. This may account for the lack of correlation between inflammation and axon loss in later stages of the disease. Thus the microenvironment of the axon may be critical to determine its capacity for long term survival. In other words, there may be a balance between the deleterious effects of inflammation and other toxic mediators and the trophic axonoprotective environment of surrounding cells. Once that balance is tipped towards favouring toxic inflammation, axon degeneration may proceed inexorably. The goals of axonoprotective treatments, therefore, may be to tip the balance in the opposite direction by restoring the trophic environment surrounding the axon.

Oligodendrocyte Influences on Axonal Survival and Neurofilament Phosphorylation

Thus, one interpretation for post-inflammatory axon loss is that the environment in which axons find themselves is no longer sufficient supportive to maintain them. Specifically the severe loss of myelin and oligodendrocytes within chronic lesions may lead to an environment in which the axon lacks necessary survival factors. Indeed there has been increasing evidence for an axonotrophic role of central nervous system myelin. Furthermore oligodendrocytes have been shown to produce soluble growth factors which influence the survival and phenotype of axons. The evidence for these phenomena will be discussed.

Myelin Has Neurotrophic Properties

Myelin has neurotrophic properties in both the peripheral and central nervous system, although mechanisms differ in these two sites (Yin *et al.* 2006). Myelin mutants have provided insights into processes by which myelin is important in axon development and maintenance. However, caution should be exercised when translating the study of myelin mutants to consider chronic progressive stages of multiple sclerosis as axons of mutant animals are exposed to abnormal myelin throughout their life.

In the peripheral nervous system, Schwann cells influence peripheral axonal structure, as hypomyelinating sciatic nerves show marked reductions in NF phosphorylation and density, with associated decrease in axonal calibre (Cole *et al.* 1994). These nerves show no changes in total NF content, suggesting Schwann cell myelin has direct influences on the process of neurofilament phosphorylation. In a similar way, mice deficient in myelin-associated glycoprotein (MAG) have reduced peripheral axonal calibres, reduced NF phosphorylation and decreased inter-NF spacing, again implying a role for myelin components in NF regulation within the peripheral nervous system (Yin *et al.* 1998).

In the central nervous system, myelin also significantly influences axon development. Central myelin is composed of several proteins including proteolipid protein (PLP) and DM20 (an alternate spliced isoform), myelin basic protein (MBP), myelin associated glycoprotein (MAG), myelin-oligodendrocyte glycoprotein (MOG) and 2',3'-cyclic nucleotide 3'-phosphodiesterase (CNPase). The absence of central myelin leads to severe disability and pronounced axonal dystrophy, as demonstrated by the *Shiverer* mouse which contains a deletion in the MBP gene and thus produces no MBP protein. This leads to the complete absence of compact CNS myelin and a phenotype of tremor and seizures with decreased life span of the mice. CNS axonal changes occur within these animals, including an increase in slow axonal transport, microtubule density and number, and altered neurofilament assembly and phosphorylation (Brady *et al.* 1999; Sanchez *et al.* 2000; Kirkpatrick *et al.* 2001). The effect appears dose dependent, since mice expressing 25% of wild-type MBP, with thin compact myelin sheaths, show an intermediate phenotype.

Mice containing other mutations in myelin proteins have been produced which show abnormalities in structure of the myelin, but nevertheless some degree of myelination. These mutants may help understand which particular components of myelin are important in the long term maintenance of axons.

PLP is localised predominantly in compact myelin and is its major structural component. Interestingly, mice lacking the Plp gene (X chromosome) are able to synthesise large amounts of myelin, which only show morphological defects in the interperiod line (the area in compact myelin where external leaflets of two adjacent plasma membranes are closely apposed). These mice do however show impairment of axonal transport mechanisms and late onset axonal degeneration with 'ovoid' formation (Griffiths *et al.* 1998b; Edgar *et al.* 2004). Interestingly, it appears that axons may become dependent on oligodendrocyte support sometime after myelination has been completed, as non-myelinated axons within optic nerves from chimeric females (which show patches of normally myelinated and non-myelinated axons) never develop axonal swellings over the expected time course. This implies a role for PLP in maintaining axons that have become oligodendrocyte dependent, although the signal triggering this putative dependency remains unknown. Further studies of chimeric females have revealed a role for PLP in early stages of axon-oligodendrocyte interactions, initiating the ensheathment and myelination of a proportion of small-diameter axons (Yool *et al.* 2001). The specific role for PLP in the central nervous system has been highlighted by a recent report in which mice were engineered to express the peripheral myelin component P_0 instead of PLP centrally (Yin *et al.* 2006). Mice lacking PLP (but expressing P_0 centrally) show a structure and periodicity of myelin which resembled peripheral myelin, yet had severe disability and degeneration of myelinated axons. Those mice expressing equal amounts of PLP and P_0 show normal phenotype and no axonal degeneration. This implies a specific role of central nervous system myelin (containing PLP) in neuroprotection of central axons, and highlights the differences between the requirements of central and peripheral axons.

Another example of the differences between central and peripheral axonal requirements comes from the study of mice lacking MAG. Axonal abnormalities are not a prominent feature within the central nervous system of these mice, whereas chronic atrophy, changes in calibre and alterations in neurofilament organization of peripheral axons occur (Li *et al.* 1994; Yin *et al.* 1998).

CNPase, a further structural molecule of myelin, has also been shown to have a role in axon-glial interactions (Lappe-Siefke *et al.* 2003). Mice lacking CNPase have morphologically normal myelin, but show delayed axonal swelling and degeneration, again suggesting that specific interactions between myelin components and the axon may determine long-term survival of the axon.

These studies suggest that the presence of structurally normal myelin may be required for long term axonal survival and that its absence will lead to chronic and slow axonal dropout. Whether this translates to the pathophysiology of axonal degeneration in chronic demyelinated lesions is unclear. Axons within myelin mutants are never exposed to structurally normal myelin, and so may develop different trophic requirements to axons ensheathed with structurally normal myelin.

As outlined, influences of myelin on neurofilament phosphorylation have been shown although the exact influence of individual myelin protein on the process are unclear. Furthermore the mechanism by which neurofilaments become dephosphorylated following demyelination is unknown.

Oligodendrocyte Derived Growth Factors Influence Axonal Survival and Neurofilament Phosphorylation

Growth factors have effects on neurofilament phosphorylation levels and there is increasing evidence that growth factors released by oligodendrocytes may influence axonal phenotype in this way. The close proximity of the oligodendrocyte to the axon would imply the possibility of cross talk between the two cell types. Indeed axon-derived factors are known to influence oligodendrocytes in a way which leads to effective functioning of glial-neuronal sub-units.

Neurons secrete factors which influence oligodendrocyte behaviour and phenotype. Specifically axons have been shown to exert a proliferative effect on oligodendrocyte precursor cells. This was first shown by transection studies on the rat neonatal optic nerve (David *et al.* 1984). Experimental transection of axons leads to a decrease in the yield of oligodendrocytes that may be cultured from the nerve. Several factors have subsequently been identified as being important in this process, including PDGF, basic FGF and members of the neuregulin family (Bogler *et al.* 1990; Canoll *et al.* 1996). Axons also promote oligodendrocyte survival (Barres and Raff, 1999). This observation is inferred from several studies of post-natal optic nerve transection . For instance, cutting the P8 optic nerve behind the eye leads to a fourfold increase in oligodendrocyte apoptosis, detected three days after transection (Barres *et al.* 1993). However, oligodendrocyte death induced by transection of the optic nerve is abrogated by locally elevated neuregulin concentrations (Fernandez *et al.* 2000). Neuregulins (Ngr1) are also crucially important in determining myelin sheath thickness and matching it to axon diameter (Michailov *et al.* 2004). Thus, signaling between neurons and cells of the oligodendrocyte lineage appears to be important in matching oligodendrocyte numbers to axonal surface area and also in ensuring that the correct cellular substrates are in place before the onset of myelination and formation of mature myelinating units.

In a similar way oligodendrocytes secrete soluble factors which influence axonal survival and phenotype. This phenomenon has been elucidated by a series of *in vitro* experiments, and there significance *in vivo* has yet to be fully identified (Du and Dreyfus, 2002). Compared to well documented and extensive studies of astrocyte-derived neurotrophic production, oligodendrocytes have traditionally been considered to be less important providers of such factors in the central nervous system.

Byravan *et al.* (1994) initially showed nerve growth factor (NGF) and brain-derived neurotrophic factor (BDNF) mRNA production by an immortalized oligodendrocyte cell line. This observation was prompted by observations of the ability of this line to promote neurite extension in PC12 cells. Furthermore, these workers showed co-localisation of NGF protein in both oligodendrocyte precursor cells and mature oligodendrocytes. A later study showed expression of NGF, BDNF and neurotrophin-3 (NT-3) mRNA in cultures of basal forebrain oligodendrocytes (Dai *et al.* 2003). Expression of these mRNAs may be altered by factors such as glutamate and potassium, suggesting that neuronal signals may provide feedback to regulate oligodendrocyte-mediated neurotrophism. Furthermore, Shinar and McMorris (1995) showed insulin-like growth factor type-1 (IGF-1) mRNA production by cultured oligodendrocytes. An extensive study of glial cell-line derived neurotrophic factor (GDNF) family member mRNA production in cells of oligodendrocyte lineage has shown GDNF mRNA production in oligodendrocyte cell line and within differentiated primary oligodendrocyte cultures (Strelau and Unsicker, 1999).

The functional significance of these findings has been shown by several studies suggesting that oligodendrocyte derived factors improve the survival of neurons and, more specifically, axons in culture. Meyer-Franke *et al.* (1995) detected an as yet uncharacterised soluble protein of >30kDa derived from mature oligodendrocytes that improved the survival of post-natal retinal ganglion cells. Later studies confirmed IGF-1 production by cells of the oligodendrocyte lineage and oligodendrocyte derived IGF-1 improves the survival of stressed neurons under conditions of trophic factor deprivation (Wilkins *et al.* 2001). Furthermore oligodendrocyte derived factors also increase levels of phosphorylated neurofilament within cultured neurons and thus aid axonal survival under the same conditions of trophic factor deprivation (Wilkins *et al.* 2003). Specifically, GDNF is the growth factor that appears to be of importance in this phenomenon and GDNF is only produced by differentiated oligodendrocytes and not precursor cells. GDNF promotes increased levels of neurofilament phosphorylation via MAPkinase/ Erk signalling pathways, as inhibition of this pathway attenuates the process and GDNF strongly activates Erk signalling within neurons. Mechanisms by which neurofilaments become phosphorylated are complex, but a role for MAPkinase /Erk has been established (Pang *et al.* 1995; Veerana *et al.* 1998; Li *et al.* 1999). Using a similar approach, oligodendrocyte derived growth factors and GDNF has also been shown to attenuate nitric oxide mediated destruction of axons, again via MAPkinase /Erk signaling (Wilkins and Compston, 2005). Again, addition of these factors significantly increases levels of phosphorylated neurofilament within cultured axons. Furthermore this study emphasises the interaction of several different signalling pathways which are involved in neuroprotective activities. p38 MAPkinase is strongly activated by addition of nitric oxide to neurons, yet inhibiton of p38 leads to activation of MAPkinase /Erk pathways. This implies a central role for Erk activation in protection of axons from inflammatory insults.

These studies have shown considerable neurotrophic and axonotrophic effects of oligodendrocytes. Both myelin and oligodendrocyte derived growth factors exert influences on long-term neuronal survival, specifically of the axonal subunit, and influence neurofilament phosphorylation within the axon.

Conclusion

Axon loss and dysfunction underlies clinical progression in multiple sclerosis. Understanding of the mechanisms involved in the process of axonopathy is vital in order to devise therapies which address specific pathophysiological processes. The concept of several mechanisms occurring within the time course of the disease probably belies the failure of current therapies which only address the inflammatory phase of the disease. Non-inflammatory processes are likely to be important in later stages of the disease and may be responsible clinical disease progression to a large degree. This 'neurodegenerative' phase of the disease is now a major target for research and in the coming years treatments will need to address this aspect. Potential treatments include remyelination therapies, delivery of growth factors or neuroprotectant drug therapies. Of these, remyelination has been the subject of extensive research. In theory, remyelination has the potential to address many aspects of disease pathology and offers the hope of recovery from symptoms. In reality, the choice of remyelinating cell, achieving satisfactory remyelination and targeting multiple lesions which are widely disseminated in space, are major obstacles preventing it becoming an imminent therapeutic option. Drugs and growth factor treatments which address axon degeneration and neurofilament dephosphorylation also seem likely to become attractive additions to the clinician's range of options available to treat patients with progressive disease. With the expansion in knowledge concerning axon pathology in multiple sclerosis such treatment options may not be far away.

References

Bagasra O, Michaels FH, Zheng YM *et al*. Activation of the inducible form of nitric oxide synthase in the brains of patients with multiple sclerosis. *Proc.Natl.Acad.Sci.U.S.A* 1995; 92: 12041-12045.

Bal-Price A, Brown GC. Inflammatory neurodegeneration mediated by nitric oxide from activated glia-inhibiting neuronal respiration, causing glutamate release and excitotoxicity. *J.Neurosci.* 2001; 21: 6480-6491.

Barres BA, Jacobson MD, Schmid R, Sendtner M, Raff MC. Does oligodendrocyte survival depend on axons? *Curr.Biol. 1993;* 3: 489-497.

Barres BA, Raff MC. Axonal control of oligodendrocyte development. *J.Cell Biol. 1999;* 147: 1123-1128.

Bitsch A, Schuchardt J, Bunkowski S, Kuhlmann T, Bruck W. Acute axonal injury in multiple sclerosis. Correlation with demyelination and inflammation. *Brain 2000;* 123 (Pt 6): 1174-1183.

Bjartmar C, Kidd G, Mork S, Rudick R, Trapp BD. Neurological disability correlates with spinal cord axonal loss and reduced N-acetyl aspartate in chronic multiple sclerosis patients. *Ann.Neurol. 2000;* 48: 893-901.

Bogler O, Wren D, Barnett SC, Land H, Noble M. Cooperation between two growth factors promotes extended self-renewal and inhibits differentiation of oligodendrocyte-type-2 astrocyte (O-2A) progenitor cells. *Proc.Natl.Acad.Sci.U.S.A 1990;* 87: 6368-6372.

Brady ST, Witt AS, Kirkpatrick LL *et al*. Formation of compact myelin is required for maturation of the axonal cytoskeleton. *J.Neurosci. 1999;* 19: 7278-7288.

Brown GC, Borutaite V. Nitric oxide inhibition of mitochondrial respiration and its role in cell death. *Free Radic.Biol.Med. 2002;* 33: 1440-1450.

Byravan S, Foster LM, Phan T, Verity AN, Campagnoni AT. Murine oligodendroglial cells express nerve growth factor. *Proc.Natl.Acad.Sci.U.S.A 1994;* 91: 8812-8816.

Canoll PD, Musacchio JM, Hardy R, Reynolds R, Marchionni MA, Salzer JL. GGF/neuregulin is a neuronal signal that promotes the proliferation and survival and inhibits the differentiation of oligodendrocyte progenitors. *Neuron 1996;* 17: 229-243.

Cole JS, Messing A, Trojanowski JQ, Lee VM. Modulation of axon diameter and neurofilaments by hypomyelinating Schwann cells in transgenic mice. *J.Neurosci. 1994;* 14: 6956-6966.

Coles AJ, Wing MG, Molyneux P *et al*. Monoclonal antibody treatment exposes three mechanisms underlying the clinical course of multiple sclerosis. *Ann.Neurol. 1999;* 46: 296-304.

Compston A, Coles A. Multiple sclerosis. *Lancet 2002;* 359: 1221-1231.

Compston A. Making progress on the natural history of multiple sclerosis. *Brain 2006;* 129: 561-563.

Confavreux C, Vukusic S, Moreau T, Adeleine P. Relapses and progression of disability in multiple sclerosis. *N.Engl.J.Med. 2000;* 343: 1430-1438.

Confavreux C, Vukusic S. Natural history of multiple sclerosis: a unifying concept. *Brain 2006;* 129: 606-616.

Craner MJ, Lo AC, Black JA, Waxman SG. Abnormal sodium channel distribution in optic nerve axons in a model of inflammatory demyelination. *Brain 2003;* 126: 1552-1561.

Craner MJ, Newcombe J, Black JA, Hartle C, Cuzner ML, Waxman SG. Molecular changes in neurons in multiple sclerosis: altered axonal expression of Nav1.2 and Nav1.6 sodium channels and Na+/Ca2+ exchanger. *Proc.Natl.Acad.Sci.U.S.A 2004;* 101: 8168-8173.

Dai X, Lercher LD, Clinton PM *et al.* The trophic role of oligodendrocytes in the basal forebrain. *J.Neurosci. 2003;* 23: 5846-5853.

David S, Miller RH, Patel R, Raff MC. Effects of neonatal transection on glial cell development in the rat optic nerve: evidence that the oligodendrocyte-type 2 astrocyte cell lineage depends on axons for its survival. *J.Neurocytol. 1984;* 13: 961-974.

Davie CA, Barker GJ, Webb S *et al.* Persistent functional deficit in multiple sclerosis and autosomal dominant cerebellar ataxia is associated with axon loss. *Brain 1995;* 118 (Pt 6): 1583-1592.

DeLuca GC, Ebers GC, Esiri MM. Axonal loss in multiple sclerosis: a pathological survey of the corticospinal and sensory tracts. *Brain 2004;* 127: 1009-1018.

DeLuca GC, Williams K, Evangelou N, Ebers GC, Esiri MM. The contribution of demyelination to axonal loss in multiple sclerosis. Brain 2006.

Du Y, Dreyfus CF. Oligodendrocytes as providers of growth factors. *J.Neurosci.Res. 2002;* 68: 647-654.

Dubois-Dalcq M, Ffrench-Constant C, Franklin RJ. Enhancing central nervous system remyelination in multiple sclerosis. *Neuron 2005;* 48: 9-12.

Edgar JM, McLaughlin M, Yool D *et al.* Oligodendroglial modulation of fast axonal transport in a mouse model of hereditary spastic paraplegia. *J.Cell Biol. 2004;* 166: 121-131.

Elhanany E, Jaffe H, Link WT, Sheeley DM, Gainer H, Pant HC. Identification of endogenously phosphorylated KSP sites in the high-molecular-weight rat neurofilament protein. *J.Neurochem. 1994;* 63: 2324-2335.

Ferguson B, Matyszak MK, Esiri MM, Perry VH. Axonal damage in acute multiple sclerosis lesions. Brain 1997; 120 (Pt 3): 393-399.

Fernandez PA, Tang DG, Cheng L, Prochiantz A, Mudge AW, Raff MC. Evidence that axon-derived neuregulin promotes oligodendrocyte survival in the developing rat optic nerve. *Neuron 2000;* 28: 81-90.

Frohman EM, Racke MK, Raine CS. Multiple sclerosis--the plaque and its pathogenesis. *N.Engl.J.Med. 2006;* 354: 942-955.

Fuchs E, Weber K. Intermediate filaments: structure, dynamics, function, and disease. *Annu.Rev.Biochem. 1994;* 63: 345-382.

Ghatan S, Larner S, Kinoshita Y *et al.* p38 MAP kinase mediates bax translocation in nitric oxide-induced apoptosis in neurons. *J.Cell Biol. 2000;* 150: 335-347.

Golde S, Chandran S, Brown GC, Compston A. Different pathways for iNOS-mediated toxicity in vitro dependent on neuronal maturation and NMDA receptor expression. *J.Neurochem. 2002;* 82: 269-282.

Goldstein ME, Sternberger NH, Sternberger LA. Phosphorylation protects neurofilaments against proteolysis. *J.Neuroimmunol. 1987;* 14: 149-160.

Grant P, Pant HC. Neurofilament protein synthesis and phosphorylation. *J.Neurocytol. 2000;* 29: 843-872.

Griffiths I, Klugmann M, Anderson T et al. Axonal swellings and degeneration in mice lacking the major proteolipid of myelin. *Science 1998;* 280: 1610-1613.

Jacobs LD, Beck RW, Simon JH et al. Intramuscular interferon beta-1a therapy initiated during a first demyelinating event in multiple sclerosis. CHAMPS Study Group. *N.Engl.J.Med. 2000*; 343: 898-904.

Kapoor R, Davies M, Blaker PA, Hall SM, Smith KJ. Blockers of sodium and calcium entry protect axons from nitric oxide-mediated degeneration. *Ann.Neurol. 2003;* 53: 174-180.

Kappos L. Effect of drugs in secondary disease progression in patients with multiple sclerosis. *Mult.Scler. 2004;* 10 Suppl 1: S46-S54.

Kirkpatrick LL, Witt AS, Payne HR, Shine HD, Brady ST. Changes in microtubule stability and density in myelin-deficient shiverer mouse CNS axons. *J.Neurosci. 2001;* 21: 2288-2297.

Kornek B, Storch MK, Weissert R et al. Multiple sclerosis and chronic autoimmune encephalomyelitis: a comparative quantitative study of axonal injury in active, inactive, and remyelinated lesions. *Am.J.Pathol. 2000;* 157: 267-276.

Kuhlmann T, Lingfeld G, Bitsch A, Schuchardt J, Bruck W. Acute axonal damage in multiple sclerosis is most extensive in early disease stages and decreases over time. *Brain 2002;* 125: 2202-2212.

Lappe-Siefke C, Goebbels S, Gravel M et al. Disruption of Cnp1 uncouples oligodendroglial functions in axonal support and myelination. *Nat.Genet. 2003;* 33: 366-374.

Leary SM, Davie CA, Parker GJ et al. 1H magnetic resonance spectroscopy of normal appearing white matter in primary progressive multiple sclerosis. *J.Neurol. 1999;* 246: 1023-1026.

Leterrier JF, Kas J, Hartwig J, Vegners R, Janmey PA. Mechanical effects of neurofilament cross-bridges. Modulation by phosphorylation, lipids, and interactions with F-actin. *J.Biol.Chem. 1996;* 271: 15687-15694.

Li BS, Veeranna, Gu J, Grant P, Pant HC. Activation of mitogen-activated protein kinases (Erk1 and Erk2) cascade results in phosphorylation of NF-M tail domains in transfected NIH 3T3 cells. *Eur.J.Biochem. 1999;* 262: 211-217.

Li C, Tropak MB, Gerlai R et al. Myelination in the absence of myelin-associated glycoprotein. *Nature 1994;* 369: 747-750.

Lovas G, Szilagyi N, Majtenyi K, Palkovits M, Komoly S. Axonal changes in chronic demyelinated cervical spinal cord plaques. *Brain 2000;* 123 (Pt 2): 308-317.

Lubetzki C, Williams A, Stankoff B. Promoting repair in multiple sclerosis: problems and prospects. Curr.Opin.Neurol. 2005; 18: 237-244.

Lucchinetti C, Bruck W, Parisi J, Scheithauer B, Rodriguez M, Lassmann H. Heterogeneity of multiple sclerosis lesions: implications for the pathogenesis of demyelination. *Ann.Neurol.* 2000; 47: 707-717.

Lucchinetti CF, Bruck W, Rodriguez M, Lassmann H. Distinct patterns of multiple sclerosis pathology indicates heterogeneity on pathogenesis. *Brain Pathol.* 1996; 6: 259-274.

Meyer-Franke A, Kaplan MR, Pfrieger FW, Barres BA. Characterization of the signaling interactions that promote the survival and growth of developing retinal ganglion cells in culture. *Neuron* 1995; 15: 805-819.

Michailov GV, Sereda MW, Brinkmann BG *et al.* Axonal neuregulin-1 regulates myelin sheath thickness. *Science* 2004; 304: 700-703.

Nixon RA, Paskevich PA, Sihag RK, Thayer CY. Phosphorylation on carboxyl terminus domains of neurofilament proteins in retinal ganglion cell neurons in vivo: influences on regional neurofilament accumulation, interneurofilament spacing, and axon caliber. *J.Cell Biol.* 1994; 126: 1031-1046.

Pang L, Sawada T, Decker SJ, Saltiel AR. Inhibition of MAP kinase kinase blocks the differentiation of PC-12 cells induced by nerve growth factor. *J.Biol.Chem.* 1995; 270: 13585-13588.

Petzold A. Neurofilament phosphoforms: surrogate markers for axonal injury, degeneration and loss. *J.Neurol.Sci.* 2005; 233: 183-198.

Petzold A, Eikelenboom MJ, Keir G *et al.* Axonal damage accumulates in the progressive phase of multiple sclerosis: three year follow up study. *J.Neurol.Neurosurg.Psychiatry* 2005; 76: 206-211.

Sanchez I, Hassinger L, Sihag RK, Cleveland DW, Mohan P, Nixon RA. Local control of neurofilament accumulation during radial growth of myelinating axons in vivo. Selective role of site-specific phosphorylation. *J.Cell Biol.* 2000; 151: 1013-1024.

Scolding N. Devic's disease and autoantibodies. Lancet Neurol. 2005; 4: 136-137.

Shinar Y, McMorris FA. Developing oligodendroglia express mRNA for insulin-like growth factor-I, a regulator of oligodendrocyte development. *J.Neurosci.Res.* 1995; 42: 516-527.

Smith EJ, Blakemore WF, McDonald WI. Central remyelination restores secure conduction. *Nature* 1979; 280: 395-396.

Smith KJ, Kapoor R, Hall SM, Davies M. Electrically active axons degenerate when exposed to nitric oxide. *Ann.Neurol.* 2001; 49: 470-476.

Smith KJ, Lassmann H. The role of nitric oxide in multiple sclerosis. *Lancet Neurol.* 2002; 1: 232-241.

Strelau J, Unsicker K. GDNF family members and their receptors: expression and functions in two oligodendroglial cell lines representing distinct stages of oligodendroglial development. *Glia* 1999; 26: 291-301.

Trapp BD, Peterson J, Ransohoff RM, Rudick R, Mork S, Bo L. Axonal transection in the lesions of multiple sclerosis. *N.Engl.J.Med.* 1998; 338: 278-285.

Veeranna, Amin ND, Ahn NG *et al.* Mitogen-activated protein kinases (Erk1,2) phosphorylate Lys-Ser-Pro (KSP) repeats in neurofilament proteins NF-H and NF-M. *J.Neurosci.* 1998; 18: 4008-4021.

Waxman SG, Craner MJ, Black JA. Na+ channel expression along axons in multiple sclerosis and its models. *Trends Pharmacol.Sci.* 2004; 25: 584-591.

Wilkins A, Chandran S, Compston A. A role for oligodendrocyte-derived IGF-1 in trophic support of cortical neurons. *Glia 2001;* 36: 48-57.

Wilkins A, Majed H, Layfield R, Compston A, Chandran S. Oligodendrocytes promote neuronal survival and axonal length by distinct intracellular mechanisms: a novel role for oligodendrocyte-derived glial cell line-derived neurotrophic factor. *J.Neurosci. 2003;* 23: 4967-4974.

Wilkins A, Compston A. Trophic factors attenuate nitric oxide mediated neuronal and axonal injury in vitro: roles and interactions of mitogen-activated protein kinase signalling pathways. *J.Neurochem. 2005;* 92: 1487-1496.

Yin X, Crawford TO, Griffin JW *et al.* Myelin-associated glycoprotein is a myelin signal that modulates the caliber of myelinated axons. *J.Neurosci. 1998;* 18: 1953-1962.

Yin X, Baek RC, Kirschner DA *et al.* Evolution of a neuroprotective function of central nervous system myelin. *J.Cell Biol. 2006;* 172: 469-478.

Yool DA, Klugmann M, McLaughlin M *et al.* Myelin proteolipid proteins promote the interaction of oligodendrocytes and axons. *J.Neurosci.Res. 2001;* 63: 151-164.

Zhao C, Fancy SP, Magy L, Urwin JE, Franklin RJ. Stem cells, progenitors and myelin repair. *J.Anat. 2005;* 207: 251-258.

In: New Research on Neurofilament Proteins
Editor: Roland K. Arlen, pp. 99-114
ISBN: 1-60021-396-0
© 2007 Nova Science Publishers, Inc.

Chapter V

The Value of Neurofilament-Immunohistochemistry for Identifying Enteric Neuron Types – Special Reference to Intrinsic Primary Afferent (Sensory) Neurons

*Axel Brehmer**

Institute of Anatomy I, University of Erlangen-Nuremberg, Erlangen, Germany

Abstract

The enteric nervous system (ENS), arranged in different nerve networks within the wall of the gastrointestinal tract, harbours numerous different neuron types which can be grouped into three main populations according to their functions. Intrinsic primary afferent neurons (IPANs) are the first links of enteric neuronal reflex arches, interneurons form ascending or descending chains of like neurons within the intestinal wall and motor neurons provide output to, e.g., the gut muscle, mucosa or blood vessels. Most of our knowledge of the neuronal components of enteric circuits, including functional evidence of IPANs, is derived from studies in the guinea-pig. An efficient method besides other techniques, to distinguish between functionally different neuron types in a given species is the evaluation of the chemical coding of enteric neurons, i.e. the immunohistochemical proof of the presence and co-existence of neuronal markers mainly within their cell bodies. However, the chemical codes of enteric neurons are species specific. During the last decade, various attempts have been undertaken to identify the functionally equivalent neurons in the human gastrointestinal tract.

---

* axel.brehmer@anatomie1.med.uni-erlangen.de; Tel: +49 9131 8522831; Fax: +49 9131 8522863

Combined tracing and immunohistochemical studies revealed projections and chemistries of putative inter- and motor neurons in the human.

In this paper, we describe the value of immunohistochemical staining for neurofilaments (NF), an important component of the cytoskeleton of subgroups of neurons, for both the identification of and the discrimination between different neuron types, which are visualized by their processal architectures, in particular the putative myenteric IPANs in the humans and pigs. These are morphological type II neurons, i.e., they are non-dendritic and multi-axonal neurons. The multiple axons of type II neurons run both circumferentially, within the plane of the myenteric plexus around the gut lumen, and vertically, penetrating the circular muscle and running to the submucosa and mucosa. We demonstrate that both calbindin, a marker for IPANs in the guinea-pig, and calcitonin gene-related peptide, a marker for IPANs in the pig and mouse, are absent from the majority of myenteric type II neurons in human. Thus, the main, and at this time the only applicable, way of identifying IPANs in the human ENS is their type II-morphology visualized through NF-immunohistochemistry.

Introduction

As early as the turn of the 19th to the 20th century, functional and structural characteristics of gut functions indicated a highly autonomous character of its nerves (Bayliss and Starling 1899; Trendelenburg 1917). The physiologist Langley (1900) divided autonomic nerves into three groups: sympathetic, parasympathetic and intestinal, and founded the concept and the term of the enteric nervous system (ENS; Langley 1921). During the following decades, many scientists moved away from this division (Brehmer et al. 1999b). The significance of enteric neurons was reduced to that of postganglionic relay stations of vegetative nerves. It was not until the 1970s that the term and the concept of this unique, enteric part of the autonomic nervous system was revived (Furness and Costa 1987).

General Structure of the Enteric Nervous System

The ENS is the nervous tissue embedded within the layers of the wall of the alimentary canal. Its main, intestinal component extends from the upper oesophageal to the internal anal sphincter. Additionally, the ENS includes nerve elements in the pancreas and the walls of both the gall bladder and bile ducts. The ENS consists of enteric neurons whose cell bodies lie within the wall of the gut, irrespective of the location of their axonal endings (some of them project outside the gut), axonal endings of extrinsic neurons (sympathetic and parasympathetic efferents as well as afferents) and enteric glial cells.

The histological arrangement of the enteric nervous tissue in the intestinal tube is that of nerve networks, the enteric plexuses. Of these, two types networks exist: one, the ganglionated plexuses (see below) consist of the perikarya of enteric neurons which are commonly grouped together as enteric ganglia. These ganglia contain different types of neurons and are not, as far as known at present, functional entities. The ganglia are interconnected by nerve strands containing axons. Two, the non-ganglionated plexuses contain axons but only few, or in some cases no nerve cell bodies. The different plexuses are connected via interconnecting strands.

Within all components of the enteric plexuses there are enteric glial cells which outnumber enteric neurons. They display cell-to-cell-coupling, have trophic and protective functions for enteric neurons, may be involved in neurotransmission and are suggested to be a link between the nervous and the immune system (Bush 2002; Cabarrocas et al. 2003; Jessen 2004; von Boyen et al. 2004; Rühl 2005).

Ganglionated plexuses are situated in two gut layers, the muscularis externa and the submucosa. Their architectures depend on species and gut region (Schabadasch 1930; Irwin 1931) and can be best investigated by using wholemounts rather than histological sections.

Tissue Samples, Wholemount Preparation, Antibodies

Human intestinal segments were derived from four patients undergoing surgery for tumors. The use of human tissues for these experiments was approved by the Ethical Committee of the University of Erlangen-Nuremberg.

Pig tissue samples were derived from animals aged between 13 and 15 weeks which were physically killed in a slaughterhouse. The European Communities Council Directive and animal welfare protocols approved by the local government were followed. Ileal segments were taken from the region ranging from about 1 m oral distance to the ileocecal orifice.

Further treatment of tissues has been described elsewhere (for pig samples see Brehmer et al. 2002b; for human samples see Brehmer et al. 2004b).

The preparation technique for routine wholemounts includes three steps starting from the inner, mucosal side. Firstly, the mucosa is scraped off the submucous layer; secondly, the submucosa is pulled off the external musculature; thirdly, strips of the circular muscle layer are pulled off the underlying longitudinal muscle layer. The submucosal wholemount (second step) contains one or few submucosal plexus (see below) whereas the myenteric plexus is situated on the inner side of the longitudinal muscle layer (third step). Due to this preparation procedure, the interconnecting strands between the different plexus are disrupted (see, e.g., Figure 2 a, b).

For incubations of pig wholemounts, the following primary antibodies were used: calcitonin gene-related peptide (CGRP: Peninsula IHC 6006, rabbit, 1:500), somatostatin (SOM: Santa Cruz Biotechnology sc-7820, goat, 1:200) and neurofilament (NF: Coulter-Immunotech 0168, mouse, 1:100). For secondary antibodies and further chemicals see Brehmer et al. (2002b).

For incubations of human wholemounts, the following primary antibodies were used: NF (Sigma N0142, mouse, 1:200), calbindin (CAB: Swant CB-38, rabbit, 1:1500) and CGRP (Peninsula, IHC 6006, rabbit, 1:500 or Progen 11189, rabbit, 1:100). For secondary antibodies and further chemicals see Brehmer et al. (2004b).

Ganglionated Plexuses

The myenteric plexus (Auerbach 1862, 1864) is embedded in the connective tissue lying between the longitudinal and the circular layer of the external gut musculature. This plexus consists of the prominent primary plexus with primary strands and myenteric ganglia, as well as the much finer secondary and tertiary strands. The primary nerve strands run both in the longitudinal (oro-anal) and in the transversal direction, most myenteric ganglia are situated at crossing points of the strands. The secondary strands are located at the inner (circular muscle) side of the primary component and run circumferentially around the gut. From these strands, fine nerve fibres enter the circular muscle layer. The tertiary strands are situated at the outer (longitudinal muscle) side of the primary plexus and run, in larger species (human, pig), parallel to the longitudinal musculature. They are regarded as the major source of innervation for the longitudinal muscle layer.

The submucosal plexus was first described by Meissner (1857). He recognized that this gut layer may be one of the regions of the human body with the highest density of nerves. Billroth (1858) observed that most components of the submucosal nerve networks were in closer proximity to the mucosa than to the (external) musculature whereas Drasch (1881) found that the submucosal plexus extends throughout the whole thickness of the submucosa with ganglia located at different levels. It is now acknowledged that small laboratory animals display a single-layerd submucosal plexus whereas larger species display two (pig) or even three (human) different submucosal plexuses. The first unequivocal description of different submucosal plexus types was given by Schabadasch (1930) in Macaques. In human, there are three submucosal plexuses which can be distinguished by their architectures, locations within the submucous layer and the chemical characteristics of their neurons (Gunn 1968; Hoyle and Burnstock 1989a, 1989b; Crowe et al. 1992; Dhatt and Buchan 1994; Wedel et al. 1999).

Enteric Neuron Types

The pioneering first approach to enable distinction of different types of neurons was that of the Russian histologist Dogiel (1896, 1899). Remarkably, he made functional suggestions based on morphological observations. In his contemporary view, all neurons had one axon and several dendrites. He described type I neurons with short dendrites, presumed motor neurons, type II neurons with dendrites leaving the ganglion of origin, presumed sensory neurons, and type III neurons with dendrites longer than those of type I but being restricted to the ganglion of origin. For the latter neurons, he made no functional suggestion. It took 80 years (see above) until the morphological classification of Dogiel was specified and extended. In the pig myenteric plexus, Stach (1981-1989; Stach et al. 2000) described up to 10 morphologically defined neuron types mainly based on the combination of their dendritic architecture combined with the direction of their axonal projection in silver impregnated wholemounts.

During the last three decades, the key advances in understanding enteric circuits and their neuronal components were made in the guinea-pig by application of numerous different methods and their combinations thereof (Costa et al. 1996; Brookes 2001; Furness 2006).

One main feature of enteric neuronal organization is that the circuitry of a number of reflexes, initiated, e.g., by mechanical or chemical stimuli, is located within the enteric plexuses, i.e., within the gut wall. Functionally, enteric neurons can be grouped into intrinsic primary afferent neurons (IPANs; briefly called 'sensory' neurons), interneurons (e.g., descending or ascending interneurons) and effector neurons (smooth muscle motor neurons, secretomotor neurons etc.).

Chemical and Morphological Codes of Neurons

The chemical coding hypothesis states that functionally different neurons display different combinations of chemical markers (transmitters, structural proteins, enzymes etc.; Furness 2006). It is also generally accepted that neurons which differ in function, differ also in their structural characteristics (Peters et al. 1991). This can be termed the morphological code.

Both are important features when trying to deduce the function from single (morphological, chemical, physiological, pharmacological) features of neurons. In turn, once morphological and chemical features of a given neuron type are known and the functional categorization thus achieved and widely accepted, its identification within a given tissue sample (e.g., for histopathological diagnostics) is much simpler when using its chemical rather than its morphological code. The first requires immunohistochemistry on routine histological sections, the latter, at least in the ENS, preparation of wholemounts and an experienced observer.

The value of combined morpho-chemical analyses to distinguish different enteric neuron types is illustrated in Figure 1. Here, applying NF-immunostaining, 3 morphological neuron types of the pig myenteric plexus - II, IV, V - which were originally described based on silver impregnated wholemounts (Stach 1981-1989) are represented. We showed that the three neuron types which appear, at first glance, structurally similar (Figure 1), display different chemical codes. Type II neurons (Figure 1a) are immunopositive for CGRP but negative for SOM, type IV neurons (Figure 1b) are positive for SOM but negative for CGRP and type V neurons (Figure 1c) are co-reactive for both CGRP and SOM (Brehmer et al. 2002 a, b; Brehmer 2006). In turn, without NF-staining and subsequent morphological analysis, it would be hard to decide whether the neurons stained for CGRP and/or SOM (Figure 1a'-c', a''-c'') represent a common (CGRP/SOM-co-reactive with variable expression of CGRP and SOM) or separate neuronal population(s).

Going one step beyond Dogiel, who mainly used dendritic shapes and lengths, Stach considered the combination of somato-dendritic morphology and the direction of axonal projections as basis for classification. Myenteric type II neurons (Figure 1a) are multi-axonal and non-dendritic, they display a combined circumferential and vertical projection pattern (Stach 1981). Type IV neurons (Figure 1b) have few short, tapering dendrites and a single axon. In myenteric plexus/longitudinal muscle wholemounts, axons of type IV neurons can typically be followed into disrupted interconnecting strands (Stach 1982). By application of DiI-tracing, we have demonstrated that both type II and type IV neurons project into mucosal villi (Brehmer et al. 1999a). Type V neurons occur in two forms, as single cells and in aggregates (Stach 1985). In this chapter, we refer only to single cells (Figure 1c). These have very few, sometimes only one, dendrite(s) which are (is) strikingly long and branched. The single axons of type V neurons, which may emerge either from the soma or a dendrite, run mostly anally within the myenteric plexus. As illustrated in Figure 1, all three types of NF-stained neurons have ovoid, smoothly contoured somata. Therefore, the visualization and charaterization of their processes is crucial for their morphological differentiation. Axons are processes which leave the ganglion of origin and display a uniform calibre for as long as they can be followed. Also after ramification, the branches of type II processes retain this axonal character. All other processes are interpreted as dendrites.

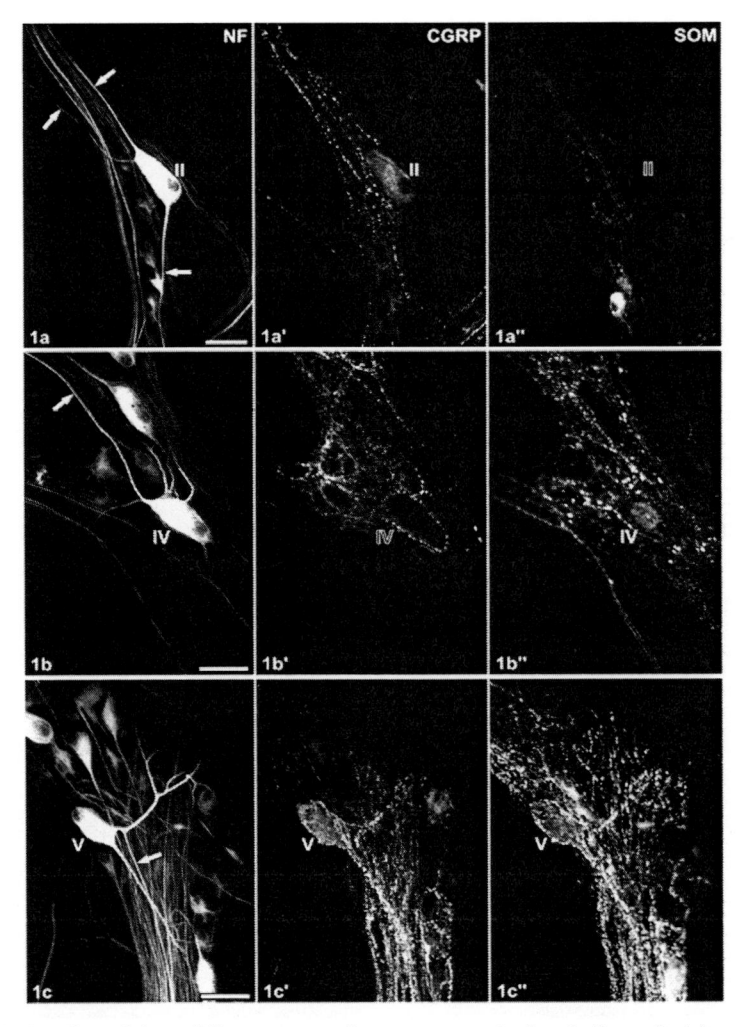

Figure 1.Chemical coding of three different myenteric neuron types in the pig ileum. In (**a**), a neurofilament (NF)-stained type II neuron (II) is depicted. Its soma is smoothly contoured and ovoid, three axons (arrows) can be followed until the margins of the picture; the neuron has no dendrites. The soma of the neuron is immuno-positive for calcitonin gene-related peptide (**a'**; CGRP; II) but immuno-negative for somatostatin (**a''**; SOM; open II). In (**b**), a NF-stained type IV neuron (IV) displaying an ovoid, smoothly contoured soma, a single axon (arrow) and several short, tapering dendrites can be seen. The neuron is negative for CGRP (**b'**; open IV) but positive for SOM (**b''**; IV). In (**c**), a NF-stained type V neuron (V) displaying an ovoid, smoothly contoured soma, a single axon (arrow) and two long, branched dendrites is depicted. The neuron displays co-immunoreactivity for both CGRP (**c'**; V) and SOM (**c''**; V).Calibration bars = 50 μm.

Although the silver impregnation technique achieved the adequate visualization of processal architecture applicable as basis for the classification of enteric neurons, the general acceptance of the Stach classification of pig enteric neurons was and is restricted (Brehmer 2006). Consequently, we tried to visualize enteric neurons by a method allowing us to simultaneously characterize them chemically. By applying of NF-immunohistochemistry, we were able to visualize enteric neurons in a quality equivalent to silver impregnation. Thus, by utilizing NF-staining as a key link between morphology and chemistry and by subsequent identification of different chemical codes of different neuron types, we corroborated the distinction of the three populations as separate neuron types which had previously solely been defined through their morphology (Stach 1981-1989).

Intrinsic Primary Afferent Neurons

The functional characterization of IPANs including their structural, chemical, electrophysiological and pharmacological features has been established in the guinea-pig. According to morphological classification, IPANs in the guinea-pig represent Dogiel type II neurons.

Based on observations in silver impregnated wholemounts of the pig myenteric plexus, Stach (1981) suggested, in contrast to Dogiel, that all processes of type II neurons are axons. Consequently, he described type II neurons as multi-axonal (some of them being 'only' pseudouni-axonal), non-dendritic neurons. Concerning functionality, Hendricks et al. (1990) later confirmed this view by showing that all processes of guinea-pig type II neurons conduct action potentials. There are two main directions of axonal projections of myenteric type II neurons: circumferentially, mainly within secondary strands of the myenteric plexus, and vertically, through interconnecting strands to the mucosa. This projection pattern has been shown in the pig (Stach 1981, Brehmer et al. 1999a), the guinea-pig (Furness et al. 1990) and the mouse (Nurgali et al. 2004; Furness et al. 2004b) and is likely also present in rat (Mann et al. 1997), human (Stach et al. 2000, Brehmer et al. 2004b) and sheep (Chiocchetti et al. 2004).

An important feature of the majority of myenteric Dogiel type II neurons in the guinea-pig is their type-specific immunoreactivity for CAB, a calcium binding protein (Furness et al. 1988, 1990). In an immunocytochemical study, Pompolo and Furness (1988) found that CAB-reactive type II neurons receive synaptic input from CAB-reactive nerve terminals. Based on these structural and a number of complementary functional data it has been concluded that Dogiel type II neurons in the guinea-pig form self-reinforcing networks (Furness et al. 1998, 2004a): a single (e.g. mucosal) stimulus results in responses in many type II neurons. They are the first neuronal links of intrinsic enteric reflexes and are therefore considered as IPANs of the guinea-pig ENS.

Besides their immunoreactivity for CAB, the multi-axonal type II neurons in the guinea-pig myenteric plexus also display characteristic electrophysiological features, one of them being the after-hyperpolarization (AH)-phenomenon: each action potential is followed by a prolonged AH for several seconds (Hirst et al. 1974; Iyer et al. 1988). In other, uni-axonal neurons with short dendrites the AH-phenomenon could not be observed. Quantitatively, the

IPANs amount to between 25 to 30% of the complete myenteric neuronal population in the guinea-pig ileum (Furness 2006), most of them being CAB-immunoreactive.

In pig, guinea-pig and human, there are also dendritic type II neurons (i.e., multi-axonal and dendritic) which may also represent IPANs (Stach 1989; Bornstein et al. 1991; Stach et al. 2000; Brehmer 2006). In the guinea-pig, they have long anal projections (Brookes et al. 1995).

Species Specifity

It is widely known that simple extrapolation of experimental data deriving from one laboratory animal to another - let alone to human - is not possible. This is also true, next to other biological features, concerning the chemical code of (enteric) neurons (Gershon et al. 1994). There are some transmitters or related substances which seem to be conserved across mammalian species whereas others vary (Furness 2006). Irrelevant of its significance, a consequence of these variations may be that the chemical coding of equivalent neurons of different species may show diversities (this chapter). In turn, non-equivalent neurons in different species may display a similar or even the same chemical code (as far as known). An example for the latter may be myenteric neurons co-immunoreactive for choline acetyl transferase and SOM. In the guinea-pig ENS, these are interneurons (Portbury et al. 1995; Song et al. 1997; Pompolo and Furness 1998), in the pig they represent secretomotor neurons (Hens et al. 2000; Brown and Timmermans 2004; Brehmer 2006).

Concerning the IPANs of the ENS, Brookes et al. (1987) already showed that the electrophysiological AH-phenomenon, typical for IPANs in the guinea-pig ileum, can only be observed in a small minority of human colonic myenteric neurons. Both in the pig (Cornelissen et al. 2000) and in the mouse (Nurgali et al. 2004), neurons identified as being morphological type II did not exhibit a marked AH-phenomenon.

Attempts of immunohistochemical characterization of the Dogiel type II neurons in other species than the guinea-pig revealed that CAB is not suitable as a marker for neurons which were suggested to be IPANs. In the pig, CGRP has been demonstrated to stain numerous Dogiel type II neurons (Scheuermann et al. 1987; Hens et al. 2000; Brehmer et al. 2002a) and has been applied as type-specific marker in an immunocytochemical study (Scheuermann et al. 1991). This finding led Timmermans et al. (1992) to suggest that the relatively few myenteric, CGRP-immunoreactive neurons in human myenteric plexus might be Dogiel type II, putative IPANs. In the mouse colon, immunoreactivity for CAB was found in Dogiel type II neurons but CGRP has been proven to be the most selective marker (Furness et al. 2004b). Also in the sheep ileum, CGRP is present in Dogiel type II neurons, putative IPANs (Chiocchetti et al. 2006).

Identification of Human IPANs

Morphological type II neurons, i.e. pseudouni- and multiaxonal neurons displaying a combined circumferential and vertical projection pattern, have been demonstrated in several species using various methods, namely in the pig (Stach 1981; Brehmer et al. 1999a; Hens et al. 2000), guinea-pig (Furness et al. 1990), rat (Mann et al. 1997) and mouse (Furness et al. 2004b; Nurgali et al. 2004). Consequently, we characterized human, myenteric type II neurons in NF-immunostained wholemounts (Brehmer et al. 2004b). In order to elucidate their chemical code, we investigated their immunoreactivities for markers which were previously shown to be more or less specific for IPANs in other species, namely CAB (guinea-pig) and CGRP (pig, mouse, sheep). We found that most human, myenteric type II neurons were negative for both markers (Figs. 2a-c, 3a-c), only a small minority displaying reactivity for CAB or CGRP (Figs. 2d, 3d). Thus, neither substance is suitable as a marker for human type II neurons. Instead, we found that the majority of human type II neurons is immunoreactive for calretinin, SOM and substance P (Brehmer et al. 2004b). Although functional evidence is lacking in human, neurons displaying equivalent morphology including their axonal projection pattern (here: type II in human and guinea-pig) are more likely equivalent than different concerning their function (here: IPAN).

Further, we pointed out, that the crude diagnosis of human myenteric neurons displaying 'smoothly contoured somata and several long processes' as Dogiel type II is insufficient. There is at least one human myenteric neuron population in the small intestine (i.e., type V) which is, at first glance, similar to type II neurons but actually differs in both architecture of processes and in chemical coding (Brehmer et al. 2004b).

Conclusion

Immunohistochemistry for NF has been proven to represent a key link between morphology and chemistry. It allowed the visualization and, in combination with other neuronal markers, the chemical definition of subpopulations of human myenteric neurons displaying quite different morphological and chemical features (Brehmer et al. 2004a, b; 2005; 2006) one of them being Dogiel type II neurons which are suggested to also be the IPANs in the human gut. The further morphological and chemical identification of human enteric neuron types by application of immunohistochemistry for NF will be of central importance for the development of histopathological diagnostics of gut diseases. Since we have shown that NFs stain only subpopulations of human myenteric neurons (Ganns et al. 2006), further structural markers have to be included in our morpho-chemical investigations.

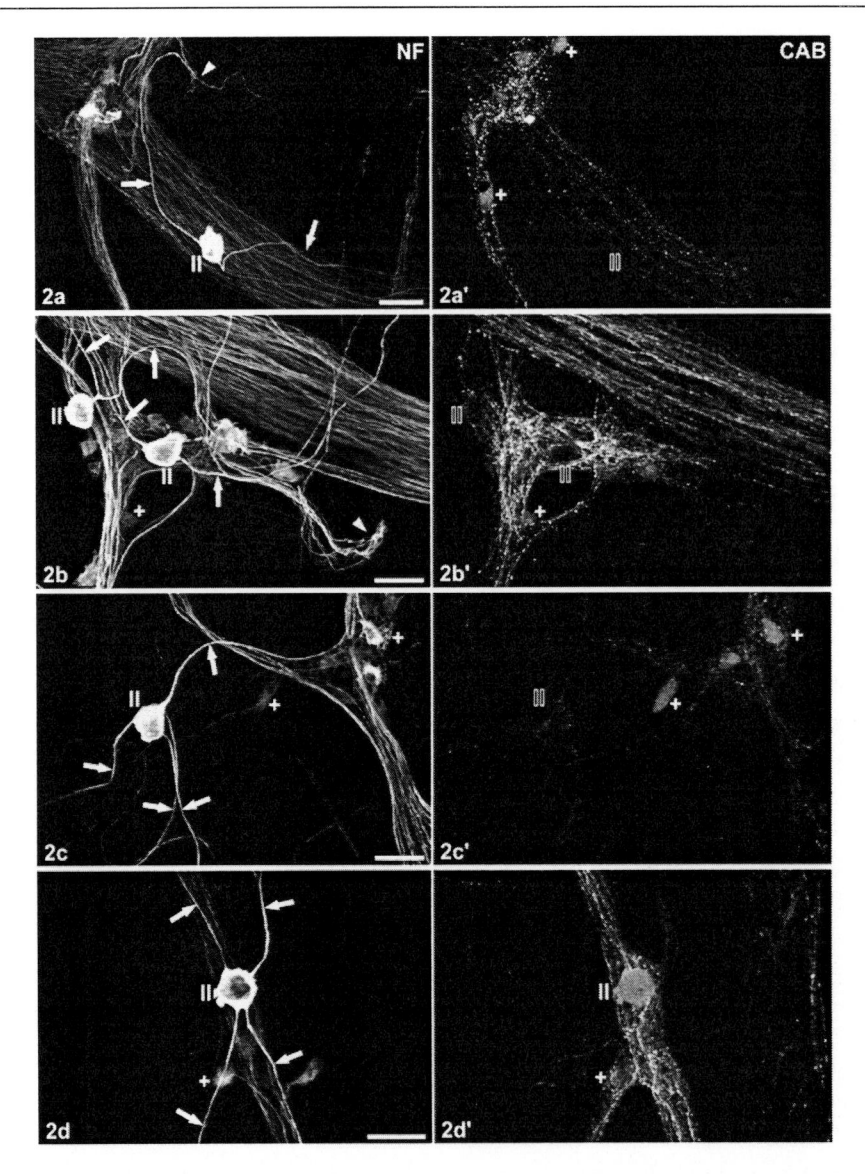

Figure 2. The majority of human type II neurons is immuno-negative for calbindin (CAB). In (a), a neurofilament (NF)-stained type II neuron (II) with two axons (arrows) is represented. One of them enters a disrupted interconnecting strand to the submucosa (arrowhead). This neuron is negative for CAB (a'; open II). In contrast, there are several small CAB-positive neurons, two of them are marked (+ in a'). These neurons are negative for NF. [Jejunum of a 40-year-old patient] In (b), two NF-stained type II neurons are marked (II). The left one is pseudouni-axonal (two arrows point at the branches after ramification), the other is bi-axonal (two arrows). One of the latter axons enters a disrupted interconnecting strand to the submucosa (arrowhead). Both neurons are negative for CAB (b'; open II). In contrast, a small neuron is weakly co-reactive for both NF and CAB (+ in b and b'). [Ileum of a 76-year-old patient] In (c), a NF-stained type II neuron (II) with four axons (arrows) is shown which is negative for CAB (c'; open II). In contrast, few small neurons are CAB-positive. Two of them, both displaying co-reactivity for CB and NF, are marked (+ in **c** and c'). [Jejunum of a 40-year-old patient] In (d), a NF-stained type II neuron (II) with four axons (arrows) is shown which is co-reactive for CAB (d'; II). Type II neurons positive for CAB represented a minority in human intestine. A small neuron (+ in d and d') is also co-reactive for CAB and NF. [Jejunum of a 40-year-old patient] Calibration bars = 50 μm.

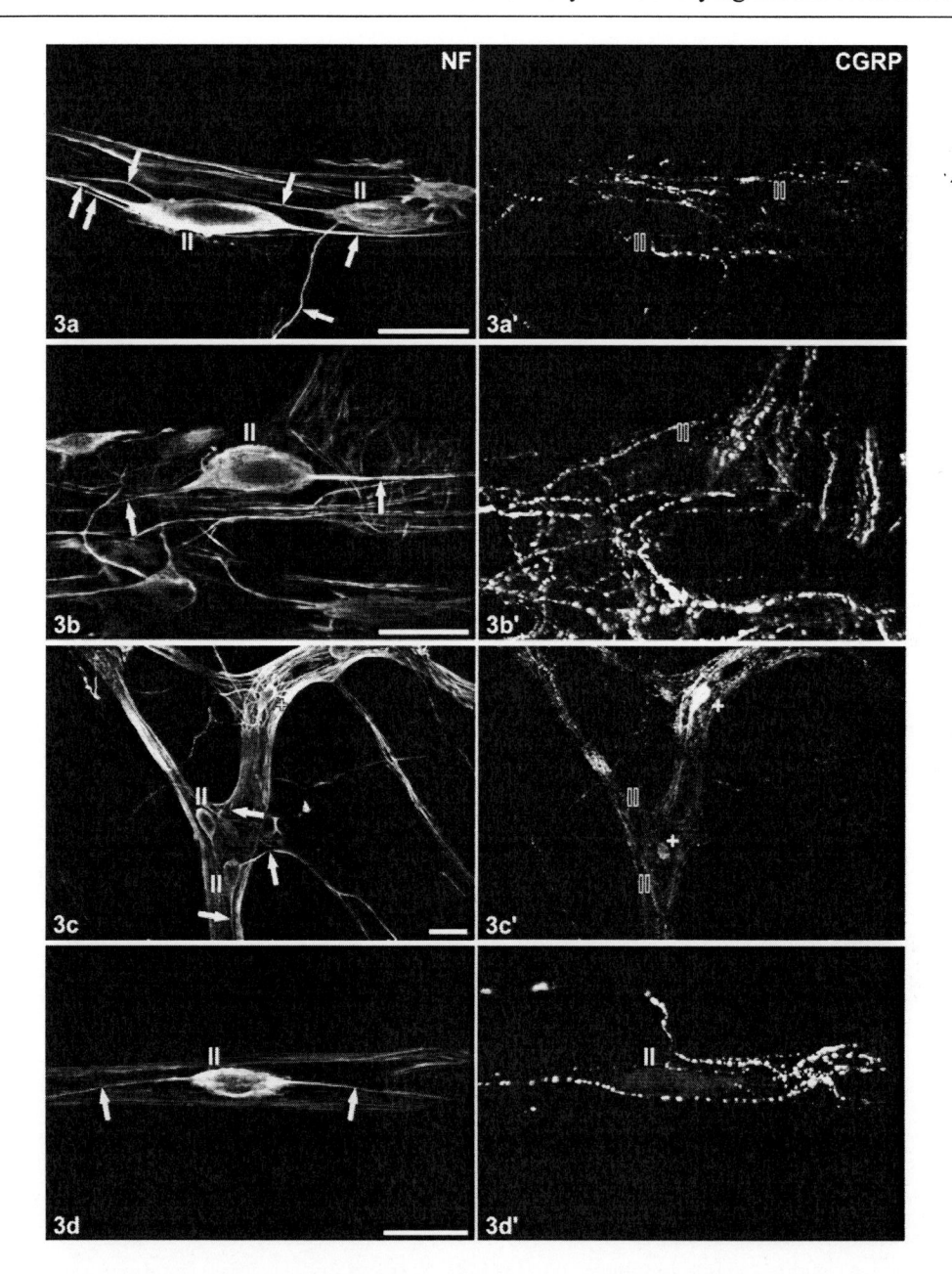

Figure 3. The majority of human type II neurons are immuno-negative for calcitonin gene-related peptide (CGRP). In (a), two neurofilament (NF)-stained type II neurons (II) are represented. One has two axons, the other four axons (arrows). Both are negative for CGRP (a'; open II). [Jejunum of a 30-year-old patient] In (b), a NF-stained type II neuron (II) with two axons (arrows) is depicted, it displays no reactivity for CGRP (b'; open II). [Jejunum of a 30-year-old patient] In (c), a ganglion with several NF-reactive neurons is represented. Two are type II neurons (II), three axons of the lower one are marked (arrows). Both type II neurons are negative for CGRP (c'; open II). In contrast, two neurons in (c') are positive for CGRP (+), the upper one is co-reactive for NF and displays a number of slender dendrites (+ in c). The shape of this neuron is strikingly different from that of the two type II neurons. [Duodenum of a 45-year-old patient] In (d), a type II neuron (II) with two axons (arrows) is seen which is weakly positive for CGRP (d'; II). Type II neurons positive for CGRP represent a minority in human intestine. [Jejunum of a 30-year-old patient] Calibration bars = 50 μm.

Acknowledgements

This and all related papers would not have been possible without those patients and body donors who consented to the use of their tissues for scientific purposes.

Furthermore, I would like to express my special thanks to Winfried Neuhuber, for both continuous support of, and interest in my work and for valuable comments on this manuscript.

I am also grateful to Karin Löschner and Stephanie Link for excellent technical assistance, to Patricia Heron for linguistic improvement of the manuscript, to Thomas Papadopoulos, Arno Dimmler, Roland Croner, Bertram Reingruber and Martin Rexer (all Erlangen), Gerhard Seitz and Barbara Blaser (Bamberg), as well as Holger Rupprecht and Daniel Ditterich (Fürth) for kind cooperation.

Work in the author's laboratory was supported by grants of the Johannes and Frieda Marohn-Stiftung (Breh/99) and the Deutsche Forschungsgemeinschaft (BR 1815/3).

References

Auerbach, L (1862). Ueber einen Plexus myentericus, einen bisher unbekannten ganglionervösen Apparat im Darmkanal der Wirbelthiere. Breslau: Morgenstern, pp 1-13.

Auerbach, L (1864). Fernere vorläufige Mittheilung über den Nervenapparat des Darmes. *Arch Pathol Anat Physiol* 30, 457-460.

Bayliss, WM; Starling, EH (1899). The movements and innervation of the small intestine. *J Physiol (Lond)* 24, 99-143.

Billroth, T (1858). Einige Beobachtungen über das ausgedehnte Vorkommen von Nervenanastomosen im Tractus intestinalis. *Arch Anat Physiol (Leipzig),* 148-158.

Bornstein, JC; Hendricks, R; Furness, JB; Trussell, DC (1991). Ramifications of the axons of AH-neurons injected with the intracellular marker biocytin in the myenteric plexus of the guinea pig small intestine. *J Comp Neurol* 314, 437-451.

Brehmer, A (2006) Structure of enteric neurons. *Adv Anat Embryol Cell Biol 186,* 1-94.

Brehmer, A; Schrödl, F; Neuhuber, W; Hens, J; Timmermans, J-P (1999a). Comparison of enteric neuronal morphology as demonstrated by DiI-tracing under different tissue handling conditions. *Anat Embryol 199,* 57-62.

Brehmer, A; Schrödl, F; Neuhuber, W (1999b). Morphological classifications of enteric neurons - 100 years after Dogiel. *Anat Embryol 200,* 125-135.

Brehmer, A; Schrödl, F; Neuhuber, W (2002a). Morphological phenotyping of enteric neurons using neurofilament immunohistochemistry renders chemical phenotyping more precise in porcine ileum. *Histochem Cell Biol 117,* 257-263.

Brehmer, A; Schrödl, F; Neuhuber, W (2002b). Correlated morphological and chemical phenotyping in myenteric type V neurons of porcine ileum. *J Comp Neurol 453,* 1-9.

Brehmer, A; Blaser, B; Seitz, G; Schrödl, F; Neuhuber, W (2004a). Pattern of lipofuscin pigmentation in nitrergic and non-nitrergic, neurofilament immunoreactive myenteric neuron types of human small intestine. *Histochem Cell Biol 121,* 13-20.

Brehmer, A; Croner, R; Dimmler, A; Papadopoulos, T; Schrödl, F; Neuhuber, W (2004b). Immunohistochemical characterization of putative primary afferent (sensory) myenteric neurons in human small intestine. *Auton Neurosci 112*, 49-59.

Brehmer, A; Lindig, TM; Schrödl, F; Neuhuber, W; Ditterich, D; Rexer, M; Rupprecht, H (2005). Morphology of enkephalin-immunoreactive myenteric neurons in the human gut. *Histochem Cell Biol 123*, 131-138.

Brehmer, A; Schrödl, F; Neuhuber, W (2006). Morphology of VIP/nNOS-immunoreactive myenteric neurons in the human gut. *Histochem Cell Biol, 125*, 557-565.

Brookes, SJH (2001). Classes of enteric nerve cells in the guinea-pig small intestine. *Anat Rec 262*, 58-70.

Brookes, SJH; Ewart, WR; Wingate, DL (1987). Intracellular recordings from myenteric neurones in the human colon. *J Physiol (Lond) 390*, 305-318.

Brookes, SJH; Song, Z-M; Ramsay, GA; Costa, M (1995). Long aboral projections of Dogiel type II, AH neurons within the myenteric plexus of the guinea pig small intestine. *J Neurosci 15*, 4013-4022.

Brown, DR; Timmermans, J-P (2004). Lessons from the porcine enteric nervous system. *Neurogastroenterol Motil 16 (Suppl 1)*, 50-54.

Bush, TG (2002). Enteric glial cells. An upstream target for induction of necrotizing enterocolitis and Crohn's disease? *Bioessays 24*, 130-140.

Cabarrocas, J; Savidge, TC; Liblau, RS (2003). Role of enteric glial cells in inflammatory bowel disease. *Glia 41*, 81-93.

Chiocchetti, R; Grandis, A; Bombardi, C; Clavenzani, P; Costerbosa, LG; Lucchi, ML; Furness, JB (2004). Characterisation of neurons expressing calbindin immunoreactivity in the ileum of the unweaned and mature sheep. *Cell Tissue Res 318*, 289-303.

Chiocchetti, R; Grandis A; Bombardi, C; Lucchi, ML; Dal Lago, DT; Bortolami, R; Furness, JB (2006) Extrinsic and intrinsic sources of calcitonin gene-related peptide immunoreactivity in the lamb ileum: a morphometric and neurochemical investigation. *Cell Tissue Res 323*, 183-196.

Cornelissen, W; De Laet, A; Kroese, ABA; Van Bogaert, P-P; Scheuermann, DW; Timmermans, J-P (2000). Electrophysiological features of morphological Dogiel type II neurons in the myenteric plexus of pig small intestine. *J Neurophysiol 84*, 102-111.

Costa, M; Brookes, SJH; Steele, PA; Gibbins, I; Burcher, E; Kandiah, CJ (1996). Neurochemical classification of myenteric neurons in the guinea-pig ileum. *Neuroscience 75*, 949-967.

Crowe, R; Kamm, MA; Burnstock, G; Lennard-Jones, JE (1992). Peptide-containing neurons in different regions of the submucous plexus of human sigmoid colon. *Gastroenterology 102*, 461-467.

Dhatt, N; Buchan, AMJ (1994). Colocalization of neuropeptides with Calbindin D_{28k} and NADPH diaphorase in the enteric nerve plexuses of normal human ileum. *Gastroenterology 197*, 680-690.

Dogiel, AS (1896). Zwei Arten sympathischer Nervenzellen. *Anat Anz 11*, 679-687.

Dogiel, AS (1899). Ueber den Bau der Ganglien in den Geflechten des Darmes und der Gallenblase des Menschen und der Säugethiere. *Arch Anat Physiol Anat Abt (Leipzig)*, 130-158.

Drasch, O (1881). Beiträge zur Kenntnis des feineren Baues des Dünndarms, insbesondere über die Nerven desselben. Sitzungsberichte der Kaiserlichen Akademie der Wissenschaften in Wien. *Abtheil III 82,* 168-198.

Furness, JB (2006). The enteric nervous system. Oxford: Blackwell.

Furness, JB; Costa, M (1987). *The enteric nervous system.* Edinburgh: Churchill Livingstone.

Furness, JB; Keast, JR; Pompolo, S; Bornstein, JC, Costa, M., Emson, PC; Lawson, DEM (1988) Immunohistochemical evidence for the presence of calcium-binding proteins in enteric neurons. *Cell Tissue Res 252,* 79-87.

Furness, JB; Trussell, DC; Pompolo, S; Bornstein, JC; Smith, TK (1990). Calbindin neurons of the guinea-pig small intestine: quantitative analysis of their numbers and projections. *Cell Tissue Res 260,* 261-272.

Furness, JB; Kunze, WAA; Bertrand, PP; Clerc, N; Bornstein, JC (1998) Intrinsic primary afferent neurons of the intestine. *Progr Neurobiol 54,* 1-18.

Furness, JB; Jones, C; Nurgali, K; Clerc, N (2004a). Intrinsic primary afferent neurons and nerve circuits within the intestine. *Progr Neurobiol 72,* 143-164.

Furness, JB; Robbins, HL; Xiao, J; Stebbing, MJ; Nurgali, K (2004b). Projections and chemistry of Dogiel type II neurons in the mouse colon. *Cell Tissue Res 317,* 1-12.

Ganns, D; Schrödl, F; Neuhuber, W; Brehmer, A (2006). Investigation of general and cytoskeletal markers to estimate numbers and proportions of neurons in the human intestine. *Histol Histopathol 21,* 41-51.

Gershon, MD; Kirchgessner, AL; Wade, PR (1994). Functional anatomy of the enteric nervous system. In: Johnson LR editor. *Physiology of the gastrointestinal tract.* Third edition. New York: Raven Press, pp 381-422.

Gunn, M (1968). Histological and histochemical observations on the myenteric and submucous plexuses of mammals. *J Anat 102,* 223-239.

Hendriks, R; Bornstein, JC; Furness, JB (1990). An electrophysiological study of the projections of putative sensory neurons within the myenteric plexus of the guinea pig ileum. *Neurosci Lett 110,* 286-290.

Hens, J; Schrödl, F; Brehmer, A; Adriaensen, D; Neuhuber, W; Scheuermann, DW; Schemann, M; Timmermans, J-P (2000). Mucosal projections of enteric neurons in the porcine small intestine. *J Comp Neurol 421,* 429-436.

Hirst, GDS; Holman, ME; Spence, I (1974). Two types of neurones in the myenteric plexus of the guinea-pig duodenum. *J Physiol (Lond) 236,* 303-326.

Hoyle, CHV; Burnstock, G (1989a). Neuronal populations in the submucous plexus of the human colon. *J Anat 166,* 7-22.

Hoyle, CHV; Burnstock, G (1989b). Galanin-like immunoreactivity in enteric neurons of the human colon. *J Anat 166,* 23-33.

Irwin, DA (1931). The anatomy of Auerbach's plexus. *Am J Anat 49,* 141-166.

Iyer, V; Bornstein, JC; Costa, M; Furness, JB; Takahashi, Y; Iwanaga, T (1988) Electrophysiology of guinea-pig myenteric neurons correlated with immunoreactivity for calcium binding proteins. *J Auton Nerv Syst 22,* 141-150.

Jessen, KR (2004) Glial cells. Int J Biochem Cell Biol 36, 1861-1867.

Langley, JN (1900). The sympathetic and other related systems of nerves. In: Schäfer EA, editor. *Text-Book of Physiology.* Edinburgh: Pentland; pp 616-696.

Langley, JN (1921). The autonomic nervous system, part 1. Cambridge: Heffer.

Mann, PT; Southwell, BR; Ding, YQ; Shigemoto, R; Mizuno, N; Furness, JB (1997). Localisation of neurokinin 3 (NK3) receptor immunoreactivity in the rat gastrointestinal tract. *Cell Tissue Res 289*, 1-9.

Meissner, G (1857). Ueber die Nerven der Darmwand. *Z Ration Med N F 8, 364*-366.

Nurgali, K; Stebbing, MJ; Furness, JB (2004). Correlation of electrophysiological and morphological characteristics of enteric neurons in the mouse colon. *J Comp Neurol 468*, 112-124.

Peters, A; Palay, SL; Webster, H deF (1991). *The fine structure of the nervous system.* New York: Oxford University Press.

Pompolo, S; Furness, JB (1988). Ultrastructure and synaptic relationships of calbindin-reactive, Dogiel type II neurons, in myenteric ganglia of guinea-pig small intestine. *J Neurocytol 17,* 771-782.

Pompolo, S; Furness, JB (1998). Quantitative analysis of inputs to somatostatin-immunoreactive descending interneurons in the myenteric plexus of the guinea-pig small intestine. *Cell Tissue Res 294,* 219-226.

Portbury, AL; Pompolo, S; Furness, JB; Stebbing, MJ; Kunze, WAA; Bornstein, JC; Hughes, S (1995). Cholinergic, somatostatin-immunoreactive interneurons in the guinea pig intestine: morphology, ultrastructure, connections and projections. *J Anat 187,* 303-321.

Rühl, A (2005). Glial cells in the gut. *Neurogastroenterol Motil 17,* 777-790.

Schabadasch, A (1930). Intramurale Nervengeflechte des Darmrohrs. *Z Zellforsch Mikrosk Anat 10,* 320-385.

Scheuermann, DW; Stach, W; De Groodt-Lasseel, MHA; Timmermans, J-P (1987). Calcitonin gene-related peptide in morphologically well-defined type II neurons of the enteric nervous system in the porcine small intestine. *Acta Anat 129,* 325-328.

Scheuermann, DW; Krammer, H-J; Timmermans, J-P; Stach, W; Adriaensen, D; De Groodt-Lasseel, MHA (1991). Fine structure of morphologically well-defined type II neurons in the enteric nervous system of the porcine small intestine revealed by immunoreactivity for calcitonin gene-related peptide. *Acta Anat 142,* 236-241.

Song, Z-M; Brookes, SJH; Ramsay, GA; Costa, M (1997). Characterization of myenteric interneurons with somatostatin immunoreactivity in the guinea-pig small intestine. *Neuroscience 80,* 907-923.

Stach, W (1981). Zur neuronalen Organisation des Plexus myentericus (Auerbach) im Schweinedünndarm. II. Typ II-Neurone. *Z Mikrosk Anat Forsch 95,* 161-182.

Stach, W (1982). Zur neuronalen Organisation des Plexus myentericus (Auerbach) im Schweinedünndarm. IV. Typ IV-Neurone. *Z Mikrosk Anat Forsch 96,* 972-994.

Stach, W (1985). Zur neuronalen Organisation des Plexus myentericus (Auerbach) im Schweinedünndarm. V. Typ V-Neurone. *Z Mikrosk Anat Forsch 99,* 562-582.

Stach, W (1989). A revised morphological classification of neurons in the enteric nervous system. In: Singer MV, Goebell H editors. *Nerves and the gastrointestinal tract.* Lancaster: Kluwer, pp 29-45.

Stach, W; Krammer, H-J; Brehmer, A (2000). Structural organization of enteric nerve cells in large mammals including man. In: Krammer H-J, Singer MV editors. *Neurogastroenterology from the basics to the clinics.* Dordrecht: Kluwer, pp 3-20.

Timmermans, J-P; Scheuermann, DW; Barbiers, M; Adriaensen, D; Stach, W; van Hee, R; De Groodt-Lasseel, MHA (1992). Calcitonin gene-related peptide-like immunoreactivity in the human small intestine. *Acta Anat 143,* 48-53.

Trendelenburg, P (1917). Physiologische und pharmakologische Versuche über die Dünndarmperistaltik. *Arch Exp Pathol Pharmakol 81,* 55-129.

von Boyen, GBT; Steinkamp, M; Reinshagen, M; Schäfer, K-H; Adler, G; Kirsch, J (2004). Proinflammatory cytokines increase glial fibrillary acidic protein expression in enteric glia. *Gut 53,* 222-228.

Wedel, T; Roblick, U; Gleiß, J; Schiedeck, T; Bruch, H-P; Kühnel, W; Krammer H-J (1999). Organization of the enteric nervous system in the human colon demonstrated by wholemount immunohistochemistry with special reference to the submucous plexus. *Ann Anat 181,* 327-337.

In: New Research on Neurofilament Proteins
Editor: Roland K. Arlen, pp. 115-130

ISBN: 1-60021-396-0
© 2007 Nova Science Publishers, Inc.

Chapter VI

Insulin Effects on Neurofilament Phosphorylation

Ruben Schechter[*] and Kenneth E. Miller

Department of Anatomy and Cell Biology, Oklahoma State University Center for Health, Science, Tulsa, OK, USA

Abstract

We demonstrated the production of neuronal insulin [I(n)] within the fetal brain in vivo and neuron cell cultures. I(n) is present within neuron cell cultures at a concentration of 320 pg/ml. We demonstrated that I(n) affects neurofilament distribution in neuron cell cultures and promotes axonal growth via mitogen activated protein kinase (MAPK or ERK).

Neurofilaments need to be phosphorylated to be distributed to the axons and form sidearms. Sidearms contribute to neurofilament anchoring, axonal caliber and growth. In diabetes mellitus, neurofilament content has been shown to be decreased. We investigated if I(n) promoted neurofilament phosphorylation and the cascade involved. Neuron cell cultures from 19 days gestational age rat brains were incubated in insulin free/serum free medium and neurofilament distribution was studied by immunohistochemistry employing a mouse monoclonal anti-phosphorylated neurofilament antibody. Neurons were treated with different inhibitors for 1 hour: 1) 10 μM of isoproterenol (an insulin receptor tyrosine kinase inhibitor); 2) 50 μM of PD98059 (a ERK inhibitor); 3) 25 μ l/ml of a guinea pig anti-porcine insulin antibody, 4) 100 μM of wortmannin a phosphatidylinositol 3 kinase (PI-3K) inhibitor. After isoproterenol or PD98059 treatments, the neurons were exposed to 5 ng/ml of insulin for 1 hour. Neurons in IFM showed phosphorylated neurofilament along the axon without interruptions. In neurons treated with isoproterenol, anti-insulin antibody or PD98059,

---

[*] Corresponding author: Reuben Schechter, rubens@chs.okstate.edu

the distribution of the phosphorylation changed to a punctuate sporadic localization along the axon. The addition of exogenous insulin to the neuron cell cultures reversed the effects of isoproterenol or PD98059, but not the anti-insulin antibody. Wortmannin did not affect neurofilament phosphorylation. Thus, I(n) has a role in neurofilament distribution and phosphorylation and, as we previously demonstrated, promoting axonal growth.

In the brain of the insulin knockout mouse (I-/-), we demonstrated that the medium molecular weight was hyperphosphorylated ($p<0.05$) when compared to wild type mice. The abnormal neurofilament hyperphosphorylation induced an aberrant assembly and transport of the neurofilament. We also quantitated kinases such as glycogen synthase kinase 3 ß (GSK-3 ß) GSK-3, ERK and c-Jun N-terminal kinase (JNK), that phosphorylate the different motifs of the neurofilament tail domain. In I-/-, we found that JNK 1 and 3 were hyperphosphorylated ($p<0.05$), ERK 1 was decreased in phosphorylation ($p<0.05$) and no changes occurred in GSK-3 when compared to wild type animals. These studies confirm that insulin via ERK phosphorylates neurofilament as shown in neuron cell cultures. Insulin deficiency may decrease the formation of medium molecular weight neurofilament sidearm, axon caliber, axonal transport velocity and, in consequence, cognition as seen today in patients of type 1 diabetes mellitus.

Introduction

In diabetes mellitus [type 1(insulin dependent) and type 2 (non-insulin dependent)], a major complication is peripheral and central neuropathy (Biessels et al, 1999). In peripheral diabetic neuropathy, in which the etiology has been more investigated, studies show neurofilament content to decrease with a loss of nerve diameter and axonal shrinkage (Fernyhough et al, 1999). Research studies of brain neuropathy are mostly descriptive and lesser effort has been dedicated to find its etiology.

Brain neuropathy is characterized by global subcortical and cortical atrophy, decrease in cognition, decrease in conduction velocity with changes in the EEG, and an increase the risk for stroke (Biessels et al, 2002). Diabetes mellitus also has been described to increase in the risk for Alzheimer's disease and other types of dementia (Arvanitakis et al, 2004). Futhermore, cerebral atrophy has been reported in the pediatric population with insulin dependent diabetes mellitus who are otherwise healthy (Sharma et al 2003; Biessels et al, 2002).

Animal research allows for the investigation of the etiology of diabetic neuropathy. Hyperphosphorylation of tau has been found within the brain of the neuronal insulin receptor knockout mouse and insulin knockout mouse (Schubert et al, 2003; Schechter et al, 2005). In the former, it was shown that neurons lack an activated Akt and, in consequence, GSK-3 ß at serine 9 is not inhibited (Schubert et al, 2003). In the insulin knockout mouse, tau hyperphosphorylation is due to activation of JNK and a reduction in the phosphorylation of MAPK (Schechter et al, 2005).

The resource of insulin in the brain may be from neuronal synthesis (Schechter et al, 1996; Devaskar et al, 1993; Deltour et al 1993) or from pancreatic beta cell production that crosses the blood brain barrier (Baskin et al, 1983). Brain endogenous insulin [I(n)] production has been shown within the neurons of fetal, newborn, and adult rats and mice by the demonstrations of insulin and insulin mRNA (Schechter et al, 1996; Devaskar et al, 1993; L. Deltour et al, 1993). I(n) is produced and secreted by neurons in fetal neuron cell cultures from rat brains incubated in a defined medium without insulin, IGF-1 or other growth factors (Schechter et al., 1998). In addition, the insulin receptor has been demonstrated within the adult and fetal brain (Schechter et al., 1996, 1994, 1998; Devaskar et al., 1993; Kenner et al., 1991; Masters et al., 1987), and the receptor is phosphorylated by insulin (Schechter et al., 1998; Masters et al., 1987; Heidenreich et al., 1989). I(n) also induces neuronal cell differentiation and axon growth (Schechter et al, 2001; Schechter et al, 1999). I(n) promotes neurofilament transport to the axon by activating ERK, especially ERK-1 (demonstrated by inhibition with PD98059, an ERK inhibitor) (Schechter et al, 1998). ERK is one of the kinases that phosphorylates the C- terminal Lys-Ser-Pro repeats of the neurofilament (Li et al, 2001; Veeranna et al, 1998).

Neurofilament distribution and content determines the caliber of the axon (Hoffman et al., 1984; Pant et al., 1995) and axonal caliber is important in establishing nerve conduction velocity (Hoffman et al., 1987; Pestronk et al., 1990). Phosphorylated neurofilaments induce the formation of the sidearm to anchor with other neurofilaments to increase the space between neurofilaments and thereby influencing axon caliber and conduction velocity (Li et al, 2001).

In peripheral diabetes neuropathy, neurofilament hyperphosphorylation occurs within the peripheral nerves (Sugimoto et al, 2000). Neurofilament hyperphosphorylation promotes reduction of protein transport and excessive concentration of neurofilament within the axons (Biessels et al, 2002). During central nervous system growth, neurofilaments are phosphorylated at a basal level to facilitate the transport of neurofilament from the neuronal body to the axons (Hornung et al, 1999).

Mitogen activated protein kinase (MAPK or ERK) and JNK belong to a family of kinases that regulate different metabolic pathways within cells (Fernyhough et al, 1999), and phosphorylate cytoskeleton proteins such as tau and neurofilament at different sites (Reynolds et al, 2000; Fernyhough et al, 1999). Insulin activates ERK 1 and 2 (Schechter et al 1998; White et al 1994) and regulates JNK (Desbois-Mouthon et al, 1998; Morino et al, 2001). Furthermore, JNK plays a role in cytoskeleton assembly and cell division (Chang et al, 2003). Factors like osmotic stress, withdrawal of growth factors, and environmental stress activate JNK (Desbois-Mouthon et al, 1998). Hyperglycemia, the major marker of diabetes mellitus, also activates JNK (Ho et al, 2000).

Central and peripheral neuropathy is the main complication in diabetes mellitus. The complication within the brain is related to alteration in cognition, brain atrophy, and risk to develop Alzheimer's type dementia (Biessels et al, 2002), but, within the brain, few studies investigate the changes in the proteins that constitute the neuron cytoskeleton, e.g., neurofilament. In this chapter, we investigate the alteration on neurofilament and alteration in the kinases, JNK, ERK, and GSK, in insulin knockout mice and in neuron cell cultures.

We undertook the current study to determine if I(n) has a role in promoting neurofilament phosphorylation during brain development. In addition, the kinase involved in I(n)'s role was investigated employing inhibitors to ERK or phosphatidylinositol-3 kinase (PI-3K). The study was performed in neuron cell cultures versus in vivo due to the difficulty of distinguishing I(n) from pancreatic insulin that crosses the blood brain barrier (Baskin et al, 1983; Wallmun et al, 1987).

Material and Methods

Reagents

Insulin 1 and insulin-2 knockout mice (I-/-) were generously given by Drs. Jami and Deltour from INSERM, Paris, France and these mice were bred to acquire the insulin knockout mice [I(-/-)]. The construction and pathology of the I(-/-) was described by Duvillie et al (Duvillie et al, 1997); diabetes mellitus developed after feeding and died between 3 to 4 days. The diagnosis of diabetes mellitus was performed by measuring glucose in urine (glucose >500 mg/dl) (kit from Sigma, St. Louis, MO).

An antibody was used to medium molecular weight phosphorylated neurofilament RMO-281 (Zymed, San Francisco,CA) that identifies phosphorylated epitopes in the C- terminal of the medium molecular weight neurofilament (Zymed, San Francisco,CA). Other neurofilament antibodies were used: SMI 311 that recognizes non-phosphorylated high molecular weight neurofilament and SMI 312 that recognizes phosphosrylated high molecular weight neurofilament (Sternberger Monoclonal Lab, Baltimore, MD).

Antibodies that recognize phosphorylated sites in different kinases were used: rabbit anti-JNK to threonine 183 and tyrosine 185 (New England Biolabs, Beverly, MA), rabbit anti-active MAPK that recognizes ERK 1 and 2 (Promega, Madison, WI), rabbit anti-GSK-3β serine 9 (New England Biolabs, Beverly, MA), mouse monoclonal anti-phosphotyrosine GSK α (tyrosine 279) and β (tyrosine 216) (UBI, Lake Placid, NY), and guinea pig anti-porcine insulin antibody was from Linco (St. Louis, MO). Other material included: nitrocellulose paper from S&S (Keene, NH), PhastSystem and gels from Pharmacia (Uppsala, Swedem), inhibitors to phosphatases (cocktail I and II) and protease inhibitor from Sigma (St. Louis, MO).

Western Blots

Neurofilament and kinases were quantitated by Western blots from five brains from I(-/-) and wild type mice [I(+/+)] that were homogenated in lysis buffer (10 mM Tris pH 7.4, 2 mM EDTA, 50 μl Triton X-100 with phosphatase inhibitor cocktail I and II and protease inhibitor). Protein concentration was measured by the bicinchoninic acid kit (BCA Protein Assay Kit, Pierce, Rockford, IL) and equal amounts of total protein were used (200 μg/ml per band) to normalize the Western blots. Gel electrophoresis of GSK, JNK, and ERK was performed using 12.5 % gels and medium molecular weight neurofilament (NF-M) in 7.5 % gels in SDS-PAGE buffer. The proteins were blotted into a nitrocellulose membrane employing the PhastSystem and identified using the ProtoBlot II AP System (Promega) employing the specific antibodies as described above (Schechter et al, 1998). Quantification of the neurofilament and the kinases was done by digitizing the Western blots and Image Tool (UTHSCSA) was employed to quantitate the protein levels.

Statistical Analysis

Statistical analysis of the phosphorylation of the medium molecular weight neurofilament and the kinases from the I(-/-) and I(+/+) were done employing the Student's test and $p<0.05$ was considered significant. Data are reported as the mean density and standard error of the mean (SEM).

Fetal Neuron Cell Cultures

Fetal neuron cell cultures were obtained from 19 day gestational age rat brain from Sprague-Dawley rat as described before (Schechter et al, 1998). The brains were dissociated using Dulbecco's Modified Eagle Medium (DMEM) (Gibco BRL, Grand Island, N.Y.) containing 2.5 µg/ml of trypsin and 1 µg/ml of deoxyribonuclease I (Worthington, Freehold, N.J.), plated at a ratio of 10^6 cells /ml of culture medium in 35 mm culture dishes, and incubated for three days in DMEM-serum medium (Schechter et al, 1990, 1994, 1998). To obtain neuron enriched cell cultures, the mixed cells were incubated in DMEM cytosine B-D arabinofuranoside for 48 hours to kill the astrocytes (Schechter et al., 1998) and subsequently were incubated in the insulin free/serum free defined medium containing DMEM/F12, 100 uM putrescine, 30 nM selenium, 20 nM progesterone and 5 µg/ml transferrin without the addition of insulin, IGF-I or other growth factors (insulin free/serum free medium) (Schechter et al, 1998). The neurons were incubated in the insulin free/serum free medium for five days and then used for the study (Schechter et al, 1998). We have demonstrated that the neurons in these cultures produce and secrete endogenous insulin (I(n)) (Schechter et al, 1988, 1990, 1994, 1998).

Purity of the neuron enriched cell cultures was confirmed by immunohistochemical staining employing mouse anti-pan neurofilament antibody, SMI-311 (Sternberger Monoclonal Lab, Baltimore, MD) and a mouse monoclonal anti-glial fibrillary acidic protein (BioGenex, San Ramon, CA) (Schechter et al, 1994, 1998, 1999).

Inhibitory Studies of Insulin

I(n)'s role in neurofilament phosphorylation was evaluated by using inhibitors to the insulin receptor, ERK, or PI-3K (Schechter et al., 1998). Furthermore, an anti-insulin antibody was used to neutralize I(n) as previously described (Schechter et al., 1998, 1999). The neurons were analyzed to study the effects of the treatment with the anti-insulin antibody or the inhibitors of neurofilament phosphorylation using immunohistochemistry. Neuron cell cultures incubated in the insulin free/defined medium (neurons incubated in this medium produce and secrete I(n)) were treated with 1) 10 µM of isoproterenol (Sigma, St Louis, MO), an insulin receptor tyrosine kinase inhibitor (Schechter et al., 1998), for 45 minutes, 2) 50 µM of PD98059 (New England Biolabs, Beverly, MA), a MAPK phosphorylation inhibitor (Schechter et al, 1998; Masters et al, 1987), for 90 minutes, or 3) 100 µM of wortmannin (Sigma), a phosphatidylinositol 3 kinase (PI-3K) inhibitor (Masters et al, 1987; Heidenreich et al, 1989), for 15 minutes as previously described (Schechter et al., 1998). The optimal time of incubation in the different inhibitors was determined in a previous study (Schechter et al, 1998). These treatments were done at five days of incubation of the neurons in the insulin free/serum free medium (Schechter et al, 1998).

I(n) action was neutralized by adding, to the insulin free/defined medium, 5 µl/ml of a guinea pig anti-porcine insulin antibody (Linco, St Louis, MO) (Schechter et al, 1998). The treatments began when the cells were incubated for 72 hours in the insulin free/defined medium (Schechter et al, 1998, 1999). The insulin antibody was added every 12 hours until cells were five days old (Schechter et al, 1998). The optimal concentration and time of incubation with the anti-insulin antibody was determined in a previous study (Schechter et al, 1998). Normal guinea pig serum was used to rule out possible guinea pig serum toxicity (Schechter et al, 1998, 1999). Normal guinea pig serum was added to the neuron cell cultures as described for the insulin antibody (Schechter et al, 1998).

The neuron cell cultures treated with the insulin antibody, isoproterenol or PD98059 also were analyzed for the reversibility of these inhibitory effects by the addition of 5 ng/ml of insulin for 45 minutes to the medium after the inhibitor treatment (Schechter et al, 1998). The optimal exogenous insulin concentration and time of incubation was determined previously (Schechter et al, 1998).

Immunohistochemistry of Neurofilament Phosphorylation

The neuron cell cultures incubated in the insulin free/defined medium and the neurons treated with the I(n) inhibitors were analyzed for the presence and localization of neurofilament phosphorylation employing a mouse monoclonal anti-pan phosphorylated neurofilament antibody and an immunofluorescent detection system. This mouse monoclonal anti-pan phosphorylated neurofilament antibody, SMI-312 (Sternberger Monoclonal Lab) recognizes the phosphorylated neurofilament (Stenberger et al, 1983; Ulfig at al, 1998). The neuron cell cultures were fixed (3.7% formaldehyde) and then treated with phosphate buffered saline (PBS) containing 0.05% Triton X-100 for 10 minutes. Neurons were incubated with a 1/20 dilution of normal goat serum in PBS for 1 hr. The neurons were incubated with SMI-312 antibody, dilution 1/400 overnight. The cells were incubated for 30 minutes in a 1/50 dilution of a goat anti-mouse antibody conjugated with fluorescein thiocyanate (Jackson Lab, West Grove, PA) for 1 hour. Control for the immunoreaction was performed by using pre-immune serum or by omitting the primary antibody. All the reactions were performed at room temperature and the washed between the incubation performed with PBS. The reaction was visualized in an Olympus microscope equipped with a Spot CCD camera.

Insulin Effects on Neurofilament Phosphorylation 121

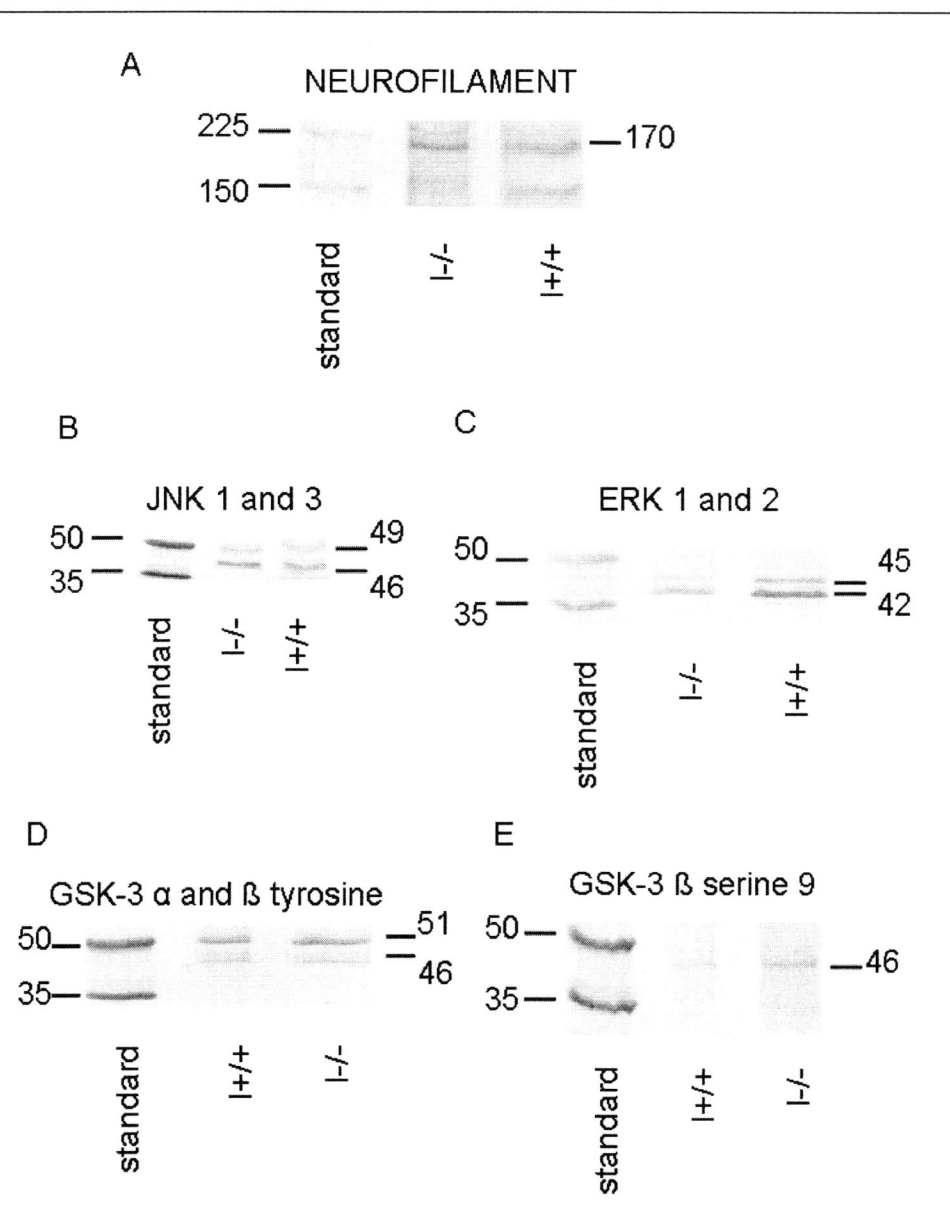

Figure 1. These Western blots correspond to the medium chain molecular weight neurofilament, JNK, ERK and GSK 3β within the I-/- and the wild type mice. A) The Western blot corresponds to the immunoreaction of the medium chain molecular weight neurofilament using an anti-phosphorylated medium chain molecular weight neurofilament. Observe the increase in phosphorylation of the medium chain molecular weight neurofilament within the I-/- ($p<0.05$). B) The Western blot corresponds to JNK immunoreaction using an anti-JNK antibody that recognizes phosphorylated JNK 1 and 3. Observe the increase in phosphorylation of both JNK isoforms within the I-/- when compared to the wild type mice ($p<0.05$). C) The Western blot corresponds to ERK immunoreaction using an anti-activated ERK antibody that recognizes ERK 1 and 2. Observe the decrease in phosphorylation in both isoforms in the I-/-, but ERK 1 was significantly decrease ($p<0.05$) when compared to the wild type mice. D) The Western blot corresponds to the phosphorylation of GSK 3 α and β tyrosine phosphorylation using an specific antibody that recognize tyrosine site in α 279 and β 216. No difference was observed between I-/- and wild type mouse ($p>0.05$). E) The Western blot corresponds to GSK 3β serine 9 phosphorylation using an anti- specific phosphorylated GSK 3β serine 9. Observe the increase in phosphorylation of the serine 9 site within the GSK 3β in the I-/-; $p<0.05$.

Result

Western Blots Analysis of the Neurofilament and Kinases within the I(-/-)

Breeding of the insulin 1 and insulin 2 knockout mice gave a 1 in 16 I(-/-). Hyperglycemia developed as soon as the animals fed with the presence of glycosuria. In the I(-/-), the urine glucose was over 500 mg/dl and, in the wild type animals, the urine glucose was below 10 mg/dl.

Medium molecular weight neurofilament was studied in this chapter because the medium molecular weight neurofilament is the most abundant neurofilament at the age of animals used (36 hour of age). The SMI 311 antibody that reacts with non-phosphorylated high molecular weight neurofilament and the antibodies to phosphorylated high molecular weight neurofilament did not show immunoreaction in the Western blots (figure 1). The specific antibody, RMO 281 recognized a band at the 170,000 molecular weight corresponding to phosphorylated medium molecular weight neurofilament. The phosphorylation of the medium molecular weight neurofilament was significantly elevated ($p<0.05$) when compared to the wild type mouse (figure 1) (wild type: 122.3 ±8.53; I(-/-) 162±10.66); showing a 30 % increase in phosphorylation.

The studies of the JNK showed a significant increase ($p<0.05$) in the phosphorylation of the isoforms JNK-1 and JNK-3 within the brain of the I(-/-) compared to wild type (JNK-1: wild type: 128.3±10.1; I(-/-): 200±1; JNK-3: wild type 108±3.5, I(-/-) 116±7.8) (figure 1). The Western blots demonstrated two bands at 46,000 molecular weight and 49,000 molecular weight corresponding to JNK-1 and JNK-3. The phosphorylation of the JNK-1 was observed to be more significant than JNK-3, with an increase in phosphorylation of 56 % for JNK-1 and as low as 7 % for JNK-3.

The studies of ERK demonstrated a significant decrease (28 %) in the phosphorylation of ERK-1 ($p<0.05$) within the I(-/-) and no significant difference in the phosphorylation of ERK-2 ($p>0.05$) when compared to wild type (ERK-1: wild type: 47.20 ±4.1; I(-/-): 34±3.3; ERK-2: wild type: 75 ±7.5; I(-/-): 54.12±6.19) (figure 1).

The analysis of GSK-3 showed no difference in the tyrosine phosphorylation in any of the GSK-3 α and β isoforms ($p>0.05$) (GSK-3 α: wild type: 148.4± 8.3; α I(-/-): 148±6.3; GSK-3β: wild type: 98.80±6.2; I(-/-): 90.20±10) (figure 1). The studies on serine 9 of GSK-3β (phosphorylation at this site inhibits the effects of the kinase) demonstrated an elevation in the phosphorylation of the site by 30 % ($p<0.05$) (wild type: 39.6 ±1.2; I(-/-): 51.40±2.46) (figure 1)

The Studies of Neurofilament Phosphorylation within the Fetal Neuron Cell Cultures

Enriched neuron cell cultures were demonstrated by immunostaining the cells with glia fibrillary acid protein and neurofilament antibody as previously described (Schechter et al, 1994, 1998, 1999).

Insulin Effects on Neurofilament Phosphorylation 123

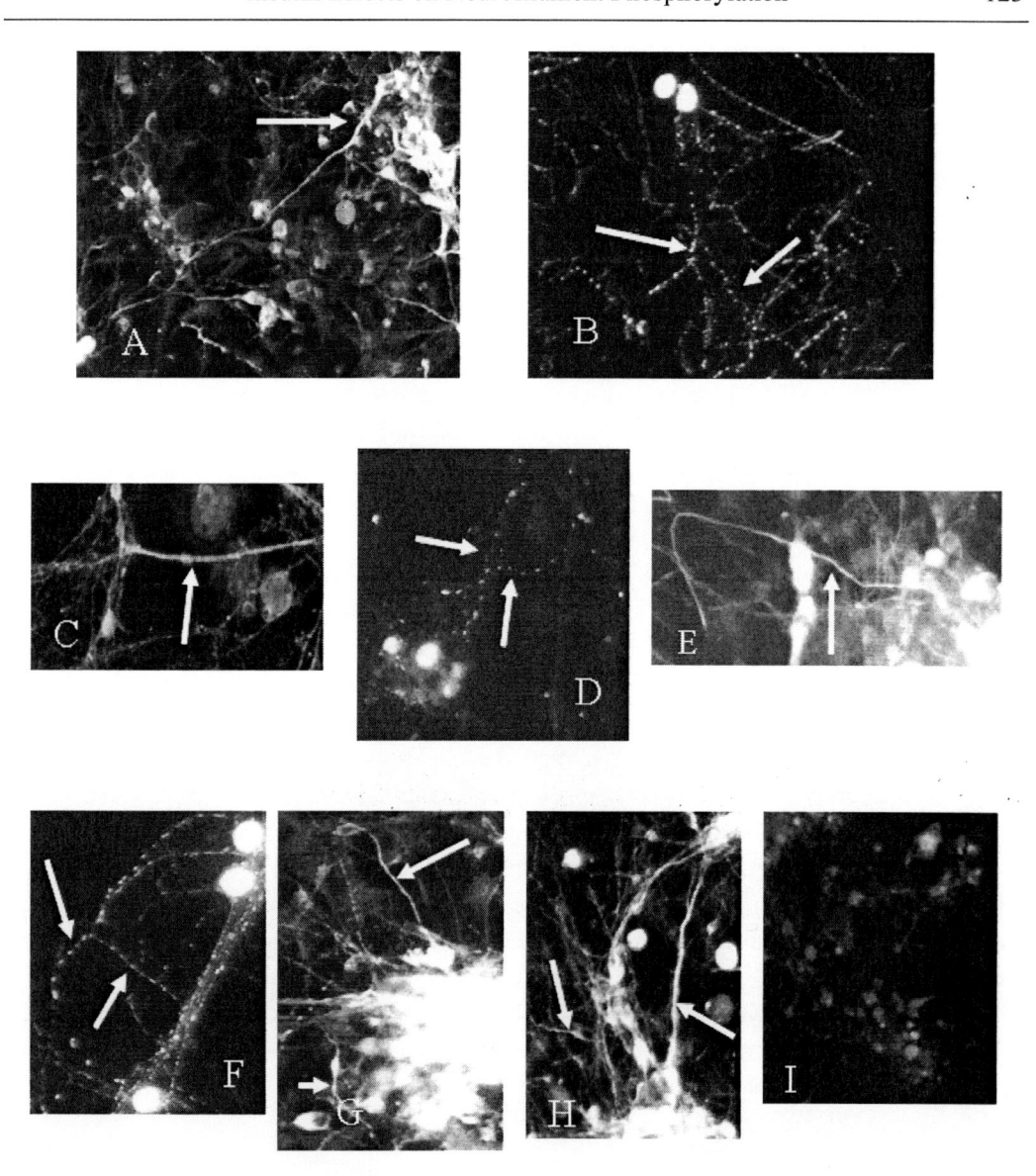

Figure 2. These microphotographs correspond to the phosphorylated neurofilament immunoreaction using SMI 312 antibody in the 19-day gestational age neuron cell cultures incubated within the insulin free/serum medium, in which the neurons produce and secrete insulin, and in the presence of inhibitors to insulin receptor, ERK, PI-3k. A) Phosphorylated neurofilament immunoreaction within neuron cell cultures incubated in the insulin free/serum free medium. Neurons in this medium produce and secrete insulin (arrows). Observe the positive phosphorylated neurofilament immunoreaction within the neurons soma and the axons (arrows). B) This neuron cell culture was incubated in the presence of five µl/ml of the guinea pig anti-porcine insulin antibody. Observe the disappearance of the phosphorylated neurofilament immunoreaction (arrows). C) This neuron cell culture was incubated in the presence of 5 µl/ml of guinea pig normal serum; no toxicity was noted. Phosphorylated neurofilament localization is within the axons and neuron soma (arrows). D) The photograph corresponds to the neuron cell cultures incubated in the presence of isoproterenol. Observe the disappearance of the diffuse neurofilament phosphorylation immunoreaction within the axon. The immunoreaction becomes punctuated within the axon (arrows). E) The photograph corresponds to neuron cell cultures incubated in the presence of exogenous insulin for 45 minutes after incubation with isoproterenol. Observe that the diffuse positive immunoreaction to the phosphorylated neurofilament is seen within the axon. F) The photograph corresponds to the neuron cell cultures incubated

in the presence of PD98059. Observe the disappearance of the diffuse neurofilament phosphorylation immunoreaction within the axon. The immunoreaction becomes punctuated within the axon (arrows). **G)** The photograph corresponds to neuron cell cultures incubated in the presence of exogenous insulin for 45 minutes after incubation with PD98059. Observe that the diffuse positive immunoreaction to the phosphorylated neurofilament is seen within the axon. **H)** The photograph corresponds to the neuron cell cultures incubated in the presence of wortmannin. Observe the diffuse neurofilament phosphorylation immunoreaction within the axon (arrows). **I)** The photograph corresponds to neuron cell cultures incubated in the presence of mouse normal serum; no immunoreaction is seen. Nucleus staining is a common finding with SMI-312. Original magnification x400.

In the fetal neuron cell cultures, growth in the insulin free/serum free medium (neurons incubated in this medium produce and secrete insulin) showed neurons with a diffuse positive neurofilament phosphorylation immunoreaction within the axon, independent of length and caliber of the axons and the soma size of the neurons (figure 2). In the axon, a reduction was evident in the neurofilament phosphorylation immunoreaction at the distal segment (figure 3). Treatment of the neuron cell cultures with wortmannin, a PI-3K inhibitor, showed no difference in the neuron localization of the neurofilament phosphorylation immunoreaction (figure 2). When the neuron cell cultures were treated with the ERK inhibitor, PD98059, the neurofilament phosphorylation immunoreaction was seen as a punctuated immunoreaction within the axon (figure 2). Treatment with isoproterenol, an insulin receptor tyrosine kinase inhibitor, produced similar immunoreaction within the axon as with PD98059. Moreover, the isoproterenol treatment resulted in the retraction of the axon and a rounding at the neuron cell body (Schechter et al, 1998). When insulin was reintroduced to the neuron cell cultures treated with PD98059 or isoproterenol, the neurofilament phosphorylation immunoreaction became diffuse again and the isoproterenol treated neurons returned to previous morphology (figure 2). When the neurons were treated with insulin antibody, the neurofilament phosphorylation immunoreaction was not detected and the cells became hypertrophic and round (Schechter et al, 1998, 1999) (figure 2). The treatment of neuron cell cultures with guinea pig normal serum did not show any difference in morphology or phosphorylated neurofilament immunoreaction localization when compared to insulin free/serum free defined medium (figure 2). The control immunostaining showed no reaction (figure 2).

Conclusion

Cognitive abnormalities and brain atrophy occur in patients with insulin dependent diabetes mellitus (Biessels et al, 2002; Sharma et al, 2003) and studies in the etiology of brain neuropathy are of clinical nature. In peripheral neuropathy, research shows that kinases like ERK and JNK hyperphosphorylate neurofilament with loss of axonal caliber (Fernyhough et al, 1999). The studies on the effect of insulin deficiency are important because neurofilaments play a critical role in the determination of axonal caliber during central nervous system growth (Grant et al, 2000). As anticipated for the developmental age, medium molecular weight neurofilament was detected in the neonatal brain tissue, but high molecular weight neurofilament was not. In addition, the medium molecular weight neurofilament was demonstrated to be hyperphosphorylated within the I-/- when compared to the wild type mice. This datum is comparable to the finding in the peripheral diabetic mellitus nervous system neuropathy, in which neurofilament is hyperphosphorylated (Fernyhough et al, 1999). The activation of the insulin receptor by insulin is known to phosphorylate a series of kinases at serine, tyrosine, or threonine sites that start a cascade that in time activate or inhibit the action of particular downstream kinases (White et al, 1994). Neurofilament is known to be phosphorylated by ERK, JNK and GSK 3β (Grant et al, 2000; Lesor et al, 1999; Reynolds et al, 2000; Fernyhough et al, 1999; Sharma et al, 2003). In our study we demonstrated that JNK-1 and JNK-3 are hyperphosphorylated in the I-/- when compared to wild type mice. In the I-/-, the phosphorylation of ERK-1 was reduced when compared to the wild type mice, but not ERK-2, which maintained similar levels of phosphorylation in the I-/- and wild type mice. ERK is known to be regulated by insulin (White et al, 1994), and, in neurons ERK, was demonstrated to be activated by insulin in neuron cell cultures (Schechter et al, 1998, 1999, 2001. Insulin in these studies regulated neurofilament transport via ERK to promote axonal growth and cell differentiation (Schechter et al, 1998, 1999, 2001). This report demonstrated that insulin regulated the activation of ERK-1 and, via ERK, the phosphorylation of neurofilament status within the neurons. These data demonstrate that insulin in vivo plays an important role in the regulation of the phosphorylation and activation of ERK-1 within the nervous system, but the inhibition of GSK 3β by activating serine 9 may be regulated by different kinases. A similar result was shown in the insulin resistant mouse in which inhibition of GSK 3β at serine 9 was present (Schubert et al, 2003).

In neurons, the phosphorylation of the tail domain of the medium molecular weight neurofilament promotes the formation of the neurofilament sidearm, inducing the anchor with other neurofilaments, increasing the space between neurofilaments, and influencing axon caliber and conduction velocity of the nerve fiber (Li et al, 2001). In neurons, aberrant accumulation of neurofilament occurs when the C terminal is phosphorylated with a decrease in neurofilament transport (Fernyhough et al, 1999). We demonstrated that the medium molecular weight was hyperphosphorylated within the brain and the phosphorylation occurred at the C terminal domain, the site where the sidearm forms. The hyperphosphorylation of the neurofilament may reduce the transport and anchor of the filament, causing a decrease in the axonal caliber and nerve conduction. This aberrant neurofilament phosphorylation may decrease the transport and assembly of the neurofilament as observed in the peripheral nervous system (Fernyhough et al, 1999). In aminals with induced diabetes mellitus, a decrease in the content of neurofilament was also demonstrated within the nerve of the peripheral nervous system (Scott et al, 1999). Moreover, abnormal increase in phosphorylation of neurofilament is detected in brain and spinal fluid of patients with Alzheimer's disease and other neurodegenerative disorders (Wang et al, 2001; Craft et al, 2004). The aberrant phophorylation of the neurofilament within the I-/- may indicate a link between diabetes mellitus and neurodegenerative disorders, e.g., Alzheimer's type dementia (Arvanitakis et al, 2004; Craft et al).

In our laboratory, we have demonstrated that insulin is synthesized and secreted by neurons in vivo and in vitro (Schechter et al, 1996, 1994, 1998, 1999) and that the insulin receptor is also localized and functional in neurons (Schechter et al, 1996, 1994, 1998; Devaskar et al, 1993; Kenner et al, 1991; Masters et al, 1987). In fetal neuron cell cultures incubated in insulin free/serum free medium, we showed that neuronal insulin is secreted in a cyclical mode with the highest concentration at 320 ng/ml (Schechter et al, 1998).

In the current chapter, we showed that neuronal insulin induces the phosphorylation of neurofilament in fetal neuron cell cultures incubated in an insulin free/serum free medium (neurons in this medium synthesized and secreted insulin [Schechter et al, 1998]) . The phosphorylated neurofilament immunoreaction was localized in the cell body and diffuse in the axon. Furthermore, the cultured neurons were treated with inhibitors to the insulin receptor (isoproterenol) and ERK (PD98059). ERK is known to phosphorylate neurofilament at sites in the C-terminal domain that induce the formation of bridges between the neurofilament and determine axon caliber. This inhibition of phosphorylation of the insulin receptor and the subsequent decrease in ERK phosphorylation showed the disappearance of the diffuse neurofilament to become punctuated along the axon. When the neuron cell cultures were treated with the anti-insulin antibody, the neurofilament phosphorylation immunoreaction disappeared and the cells became hypertrophic. In earlier studies in neuron cell cultures incubated in insulin free/serum free medium, we demonstrated that non-phosphorylated neurofilament distribution to the axon was abolished when using the same insulin receptor and ERK inhibitors and an anti-insulin antibody (Schechter et al, 1998). The PI-3K inhibitor, wortmannin, showed no effect in the localization and distribution of the neurofilament phosphorylation immunoreaction within the neuron cell cultures as with the non-phosphorylated neurofilament (Schechter et al, 1998). In current chapter, we showed that neuronal insulin participated in the regulation of ERK not only in the neuron cell cultures, but also in vivo. In the I-/- , ERK phosphorylation was decreased and, in the neuron cell cultures, ERK is not activated in presence of isoproterenol (Schechter et al, 1998, Schechter et al., 2005). The current and prior data demonstrate that neuronal insulin via ERK participates in the distribution and phosphorylation of neurofilament. ERK was described to participate in the phosphorylation of the neurofilament by activation the C- terminal domain of the neurofilament at the Lysine-Serine-Proline sites (Veeranna et al, 1998). Moreover, ERK was demonstrated to be activated by insulin in NIH 3T3 cells and Chinese Hamster ovary cells (White et al, 1994).

Neuronal insulin was described to participate in the differentiation of neurons (Schechter et al, 2001) and also induce axonal growth (Schechter et al, 1999). Neurofilament distribution to the axon induces axonal growth (Hoffman et al, 1984) and the neurofilament needs to be phosphorylated to be transported within the axons (Pant et al, 1995; Pestronk et al, 1990). Neurofilament content in the axon is important because it determines the axonal caliber (Hoffman et al, 1984) and neurofilament content establishes nerve conduction velocity (Hoffman et al, 1987). The role of neurofilament in axonal growth has been demonstrated in neurofilament knockout mice in which the nerve conduction velocity and axonal caliber was diminished (Yagihashi et al, 1990, Zhu et al, 1997, Eyer et al, 1994).

In this chapter, we presented evidence for an insulin function within the brain in the phosphorylation equilibrium of the neurofilament balance and demonstrated that deficiency of insulin interrupts this equilibrium. We theorize that pancreatic synthesized insulin, that has secretion induced by glucose level in blood and has a short half life may not be the source to maintain the constant neurofilament phosphorylation balance. Neurons are in need of a continuous source of insulin. The source of continuous insulin secretion may be from brain neuronal and choroid plexus production and secretion as shown previously (Schechter et al, 1996, 1994, 1998, 1999; Lamotte 2004).

In summary, lack of insulin (such as diabetes mellitus) induces neurofilament hyperphosphorylation by increasing the activity of JNK. These alterations in the neurofilament reduce the transport of neurofilament to the axons and decrease nerve conduction velocity (Yagihashi et al, 1990). Furthermore, the data described by us and others indicate the possibility that nervous system alterations may depend on the deficiency of insulin (Sugimoto et al, 2000). In addition, our data suggest that neuronal insulin induces neurofilament distribution and phosphorylation and that this action is dependent via ERK. These and earlier data imply a role of neuronal insulin in axon growth and neurofilament transport to the axon during brain development.

Reference

Arvanitakis Z., Wilso R.S., Bienias J.L., Evans D.A., and Bennet D.A., Diabetes mellitus and risk of Alzheimer disease and decline in cognitive function. *Arch Neurology 61 (2004)* 661-666.

Baskin D.G., Woods S., West D., van Houten M., Posner B.I., Dorsa D., and Porte Jr D., Immunocytochemical detection of insulin in rat hypothalamus and its possible uptake from cerebrospinal fluid. *Endocrinology 183 (1983)* 1818-1825.

Biessels GS, Cristino N.A., Rutten G.J., Hamers F.P.T., Erkelens D.W. and Gispen W.H., Neurophysiological changes in the central and peripheral nervous system of streptozotocin diabetic rats. Course of development and effects of insulin treatment. *Brain 122 (1999)* 757-768.

Biessel G.J., van der Heide L.P., Kama A., Bleys R.L.A.W., and Gispen W.H., Ageing and diabetes: implication for brain function. *European Journal of Pharmacology 441(2002)* 1-14.

Chang L., Jones Y., Ellisman M.H., Goldstein L.S.B., and Karin M., JNK 1 is required for maintenance of neuronal microtubules and controls phosphorylation of microtubule-associated proteins. *Developmental Cell 4 (2003)* 521-533.

Craft S. and Watson S., Insulin and neurodegenerative disease: shared and specific mechanisms. *Lancet Neurology 3 (2004)* 169-178.

Deltour L., Leduque P., Blume N., Madsen O., Dubois P., Jami J., and Bucchini D., Differential expression of the two nonallelic proinsulin genes in the developing mouse embryo. *Proc. Natl. Acad. Sci. USA 90 (1993)* 527-531.

Desbois-Mouthon C., Blivet-van Eggelpoel M.J., Auclair M., Cherqui G., Capeau J., and Caron M., Insulin differentially regulates SAPKs/JNKs and ERKs in CHO cells overexpressing human insulin receptors. *Biochemical and Biophysical Research Communications 243 (1998)* 765-770.

Devaskar S.U., Singh B.S., Carnaghi L.R., Rajakumar P.A., and Giddings S.J.. Insulin II gene expression in rat central nervous system. *Regulatory Peptide 48 (1993)* 55-63.

Duvillie B., Cordonnier N., Deltour L., Dandoy-Dron F., Itier Monthioux J.M., Jami J., Joshi R.L., and Bucchini D, Phenotypic alterations in insulin-deficient mutant mice. *Proc. Natl. Acad. Sci USA 94 (1997)* 5137-5140.

Eyer J., Peterson A.. Neurofilament-deficient axons and perikaryal aggregates in viable transgenic mice expressing a neurofilament-beta-galactosidase fusion protien. *Neuron 12 (1994)* 389-405.

Fernyhough P., Gallaghee G., Averill S.A., Priestley J.V., Hounsom L., Pate J. l, and Tomlinson D.R., Aberrant neurofilament phosphorylation in sensory neurons of rats with diabetic neuropathy. *Diabetes 48 (1999)* 881-889.

P. Grant and H.C. Pant, Neurofilament protein synthesis and phosphorylation. *J Neurocytology 29 (2000)* 843-872.

Heidenreich K.A., Toledo S.P.. Insulin receptor mediate growth effects in cultured fetal neurons: II. Activation of a protein kinase that phosphorylates ribosomal protein S6. *Endocrinology 125 (1989)* 1458-1463.

Ho F.M., Liu S.H., Liau C.S., Huang P.J., Lin-Shiau S.Y., High glucose-induced apoptosis in human endothelial cells is mediated by sequential activations of c-JUN NH_2- terminal kinase and Caspase-3. *Circulation 101 (2000)* 2618-2624.

Hoffman P.N., Griffin J.W., Price D.L.. Control of axonal caliber by neurofilament transport. *J Cell Biol 99 (1984)*, 705-714.

Hoffman P.N., Cleveland D.W., Griffin J.W., Landes P.W., Cowan N.J., Price D.L.. (1987). Neurofilament gene expression: major determinant of axonal caliber. *Proc. Natl.Acad.Sci.USA 85 (1987)* 3472-3476.

Hornung J.P. and Reiderer B.M., Medium sized neurofilament protein related to maturation of a subset of cortical neurons. *Journal of Comparative Neurology 414 (1999)* 348-360.

Kenner K.A., and Heidenreich K.A.. Insulin and insulin-like growth factors stimulate in vivo receptor autophosphorylation and tyrosine phosphorylation of a 70K substrate in cultured fetal chick neurons. *Endocrinology 139 (1991)* 301-331.

Lamotte L, Jackerott M., Bucchini D., Jami J., Joshi R.J. and Deltour L., Knock-in of diphtheria toxin A chain gene at Ins2 locus: effects on islet development and localization of Ins2 expression in the brain. *Transgenic Research 13 (2004)* 463-473.

Lesor M., Jope R.S., and Johnson G.V., Insulin transiently increases tau phosphorylation: involvement of glycogen synthase kinase-3β and Fyn tyrosine kinase. *J Neurochemistry 72 (1999)* 576-584.

Li B.S., Daniels M.P., and Pant H.C., Integrins stimulate phosphorylation of neurofilament NF-M subunit KSP repeats through activation of extracellular regulated-kinases (ERK1/ERK2) in cultured motoneurons and transfected NIH 3T3 cells. *J Neurochemistry 76 (2001)* 703-710.

Masters B.A., Shemer J., Judkins J.H., Clarke D.W., Le Roith D., and Raizada M.K.. (1987). Insulin receptors and insulin action in dissociated brain cells. *Brain Res 417 (1987)* 247-256.

Morino K., Maegawa H., Fujita T., Takahara N., Egawa K., Kashiwagi A., and Kikkawa R., Insulin-induced c-JUN N-terminal kinase activation is negatively regulated by protein kinase C. *Endocrinology 142 (2001)* 2669-2676.

Pant H.C. and Veeranna. Neurofilament phosphorylation. *Biochem Cell Biol 73 (1995)* 575-591.

Pestronk A., Watson D.F., and Yuan C.M.. Neurofilament phosphorylation in peripheral nerve: changes with axonal length and growth state. *J. of Neurochem 54 (1990)* 977-982.

Reynolds C.H., Betts J.C., Blackstock W.P., Nebreda A.R., and Anderton B.H., Phosphorylation sites on tau identified by nanoelectrospray mass spectrometry: differences in vitro between the mitogen-activated protein kinase ERK2, c-Jun N-terminal kinase and P38, and glycogen synthase kinase-3β. *J Neurochemistry 74 (2000)* 1574-1595.

Schechter R, Holtzclaw L, Sadiq F, Kahn A, and Devaskar S. Insulin systhesis by isolated rabbit neurons. *Endocrinology 123 (1988)* 505-513.

Schechter R, Sadiq HF and Devaskar SU. Insulin and insulin mRNA are detected in neuronal cell cultures maintained in an insulin-free/serum-free medium. *J Histochem Cytochem 38(6)* (1990):829.

Schechter R., Whitmire J., Wheet G.S., Beju D., Jackson K.W., Harlow R., and Gavin III J.R.. Immunohistochemical and in situ hybridization study of an insulin-like substance in fetal neuron cell cultures. *Brain Res 636 (1994)* 9-27.

Schechter R., Beju D., Gaffne T., F. Schaefer F., and Whetsell L., Preproinsulin I and II mRNA and insulin electron microscopic immunoreaction are present within the rat fetal nervous system. *Brain Research 736 (1996)*16-27.

Schechter R., Yanovitch T., Abboud M., Johnson III G., and Gaskin J., Effects of brain endogenous insulin on neurofilament and MAPK in fetal rat neuron cell cultures. *Brain Res 808 (1998)* 270-278.

Schechter R., Abboud M., Johnson III G.. Brain endogenous insulin effects on neurites growth within fetal rat neuron cell cultures. *Developmental Brain Research 116 (1999)* 159-167.

R. Schechter and M. Abboud, Neuronal synthesized insulin roles on neural differentiation within fetal rat neuron cell cultures. *Dev Brain Res 127 (2001)* 41-49.

Schechter R., Beju D., Miller K.E.. The effect of insulin deficiency on tau and neurofilament in the insulin knockout mouse. *Biochem and Biophys Resear Comm 334 (2005)* 979-986

Schubert M., Brazil D.P., Burks D.J., Kushner J.A., Ye J., Flint C.L., Farhang-Fallah J., Dikkes P., Warot X.M., Rio C., Corfas G., and White M.F., Insulin receptor-2 deficiency impairs brain growth and promotes tau phosphorylation. *J Neuroscience 23 (2003)* 7084-7092.

Scott J.N., Clark A.W., and Zochodne D.W., Neurofilament and tubulin gene expression in progressive experimental diabetes. Failure of synthesis and export by sensory neurons. *Brain 122 (1999)* 2109-2117.

Sharma J., Bakshi R., Lee D., Hachinski V., and Cha R.K.T., Cerebral atrophy in young, otherwise healthy patients with type 1 diabetes mellitus. *Neurology [suppl] 35 (2003)* 3-29.

Shubert M., Gautam D., Surjo D., Ueki K., Baudle S., Schubert D., Kondo T., Alber J., Galldiks N., Kusterman E., Arnd S., Jacobs A.H., Krone W., Kahn C.R, and Bruning J.C., Role for neuronal insulin resistance in neurodegenerative diseases. *Proc.Natl.Acad. Sci. USA 101 (2004)* 3100-3105.

Sternberger L.A., Sternberger N.H.. Monoclonal antibodies distinguish phosphorylated and nonphosphorylated forms of neurofilaments in situ. *Proc Nat Acad Sci 80 (1983)* 6126-6130.

Sugimoto K., Murakawa Y., Zhang W., Xu G., Sima A.A. F.. (2000). Insulin receptor in rat peripheral nerve: is localization and alternative sliced isoforms. *Diabetes Metab. Res.Rev. 16 (2000) 354*-363.

Ulfig N, Nickel J, Bohl J. Monoclonal antibody SMI 311 and 312 as tools to investigate the maturation of nerve cells and axonal patterns. *Cell tissue Research 291 (1998)* 433-443.

Veeranna, Amin N.D., Ahn N.G., Jaffe H., Winters C.A., Grant P., Pant H.C.. Mitogen-activated protien kinase (ERK-1, 2) phosphorylates Lys-Ser-Pro (KSP) repeats in neurofilament protein NF-H and NF-M. *J. of Neuroscience 18 (1998),* 4008-4021.

Wallum B.J., Taborsky J., Jr., Porte D., Jr., Figlewicz D., Jacobson L., Beard J., Ward W., and Dorsa D.. (1987). Cerebrospinal fluid insulin levels increase during infusions in man. *J. Clin Endocrinol Metab 64 (1987)* 190-194.

Wang J.Z., Tung Y.C., Wang Y., Li X.T., Iqbal K., and Grunke-Iqbal I., Hyper phosphorylation and accumulation of neurofilament proteins in Alzheimer disease brain and in okadaic acid-treated SY5Y cells. *FEBS Letters 507 (2001)* 81-87.

White M.R., Kahn C.R.. The insulin signaling system. *J Biol Chem 269 (1994)* 1-4.

Yagihashi S., Kamijo M., and Watanabe K.. Reduced myelinated fiber size correlates with loss of chronically streptozotocin diabets rats. *Am J Patol 136 (1990),* 1365-1373.

Zhu Q., Couillard-Depres S., Julien J.P.. Delayed maturation of regenerating myelinated axons in mice lacking neurofilament. *Exp. Neurol. 148 (1997)*299-316.

In: New Research on Neurofilament Proteins
Editor: Roland K. Arlen, pp. 131-162

ISBN: 1-60021-396-0
© 2007 Nova Science Publishers, Inc.

Chapter VII

Neurofilaments in the Mammalian Visual System: As Revealed by SMI 32 Immunohistochemistry

Zsolt B. Baldauf[*]

Neurobiology Research Group, Hungarian Academy of Sciences at the Department of Anatomy, Histology and Embryology, School of Medicine, Semmelweis University, Budapest, Hungary

Abstract

The histochemical detection of neurofilaments (Nfs) has proved to be a powerful tool in demarcating several cortical and subcortical areas in the mammalian visual system. Apart from the Nfs' cortical parcellating power, it delineates previously unequivocal histological boundaries within the associative thalamic and pretectal nuclear complexes. In addition, the accumulation of Nfs is a true marker of large brainstem oculomotor neurons and cortical layer V giant pyramidal neurons alike in various regions. The parallelism of the visual system can easily be detected and observed in the characteristic distribution of Nfs. These polypeptides are strongly expressed alongside a visual stream, the magnocellular (M) pathway, at both cortical and subcortical levels. The faster-conducting M pathway's axons are of a larger caliber, and thus display more cytoskeletal Nfs.

Research on Nfs has gained medical importance lately due to their presumptive role in various neurodegenerative diseases. A monoclonal antibody (SMI 32) readily recognizes the presence of the Nf triplet in mammalian, and also in marsupial, nervous systems. Here, the immunohistochemical distribution of SMI 32-reactive Nfs along the rodent, carnivore and primate visual system is discussed.

---

[*] Present address: Department of Anatomy, Szent-Györgyi Albert School of Medicine, University of Szeged, 40 Av. Kossuth Lajos, Szeged 6724, Hungary E-mail address: zsolt.baldauf@yahoo.com

Abbreviations

A	lamina A (feline dLGN)
A1	lamina A1 (feline dLGN)
AOS	accessory optic system
APNc	anterior pretectal nucleus, pars compacta
APNR	anterior pretectal nucleus, pars reticularis
CO	cytochrome oxidase histochemistry
C_M	lamina C, pars magnocellularis (feline dLGN)
C_P	lamina C, pars parvocellularis (feline dLGN)
dLGN	lateral geniculate nucleus, pars dorsalis
DTN	dorsal terminal nucleus
H	habenula
HTh	hypothalamus
IGL	intergeniculate leaflet
-ir	-immunoreactive
LP	lateral posterior nucleus (rodent, feline)
LPcl	lateral posterior nucleus, pars caudalis, lateral division (rodent)
LPcm	lateral posterior nucleus, pars caudalis, medial division (rodent)
LPl	lateral posterior nucleus, pars lateralis (feline)
LPm	lateral posterior nucleus, pars medialis (feline)
LPP	lateral posterior-pulvinar complex (feline)
LTN	lateral terminal nucleus
MD	monocular deprivation
MIN	medial intralaminar nucleus
MPN	medial pretectal nucleus
MT	medial temporal cortical area
MTN	medial terminal nucleus
NOT	nucleus of the optic tract
nIII	oculomotor nucleus
OPN	olivary pretectal nucleus
OT	optic tract
PGN	perigeniculate nucleus
PLLS	posterolateral lateral suprasylvian cortex
PMLS	posteromedial lateral suprasylvian cortex
PPN	posterior pretectal nucleus
PPULN	peripulvinar nucleus
PT	pretectum
PUL	pulvinar nucleus
SC	superior colliculus
SCN	suprachiasmatic nucleus
SGN	suprageniculate nucleus
SGI	stratum griseum intermedium
SGP	stratum griseum profundum

SGS   stratum griseum superficiale
SZ   stratum zonale
TRN   thalamic reticular nucleus
VB   ventrobasal nucleus
vLGN   lateral geniculate nucleus, pars ventralis

Introduction

At first, the histology of the nervous system merely consisted of the detection of cell bodies and fibers based upon their chemical affinity. Today various immunohistochemical and molecular biological techniques are available to explore the anatomical organization of the nervous system. On the one hand, these methods specifically label certain subsets of neurons and glial elements, so that they can delineate otherwise homogenous structures. On the other hand, these data also provide useful information about the functions of various cell types. The differences among neurons can be determined not only by the multifarious structural features (e.g. soma size, axon caliber and length, myelin thickness, calcium binding protein content, dendritic branching pattern, synaptic contacts, perineuronal net), but by a long cohort of electro- and neurochemical and genetic properties, such as dissimilar firing properties, membrane excitability, transmitter content, receptor and ion channel pool, transcription factor and gene expressions as well. Moreover, in the hope of detecting further anatomical or functional dissimilarity, the visualization of intracellular neurofilaments (Nfs) has attracted keen attention in the past two decades.

Apart from the emerging recognition of the role of myosins in the neurons' internal scaffolding [16], the Nfs are well-established constituents of the dynamic neuronal cytoskeleton [103, 125]. Their cellular specificity classifies them as Type IV filaments [41]. Since their size (10 nm) is between the larger filaments or microtubules (24 nm) and the smaller filaments or microfilaments (6-8 nm), the Nfs are pigeonholed as intermediate filaments. The heavy (NfH; 190-210 kDa depending on its phosphorylation state, since it is the most extensively phosphorylated protein in the brain resulting in marked changes in its molecular weight), the medium (NfM; 150 kDa) and the light (NfL; 68 kDa) subunits form the filamentous Nf triplet phosphoprotein [113, 84]. This nomenclature was originally suggested by Shaw and his coworkers in 1984 and it is still in use today [112]. Following transcription and protein composition in the cell body, the Nf subunits are transported into axons with the help of kinesin and dynein [114, 97]. Here the subunits (NfH, NfM, NfL) assemble to the functional heteropolymer triplet, which is a very stable intracellular tube system endowing the cell with a firm skeleton. The axon terminals are void of Nfs because they undergo an enzymatic degradation here [80, 81], while intact Nfs can be found in dendrites and somata. In fact, the phosphorylation of Nf at the head domain plays a critical role in inhibiting Nf assembly in the perikaryon, whereas extensive phosphorylation of the Nf tail domain taking place in axons results in the self assembly of Nf triplets [91, 135]. In addition, the Nf triplets are used to determine whether a neuron is mature, since Nf phosphorylation and assembly is generally considered specific to the adult nervous system [86].

These ternary structural polypeptides allow neurons to establish and maintain the incredibly complex and asymmetrical geometrical shapes of cells, to enhance their structural integrity and organelle motility, and to add tensile strength to neurons. Axonal myelination is controlled by Nfs [39] and in turn, myelin may influence the Nfs phosphorylation in axons [84]. Although probably not managing longitudinal axonal growth, the Nfs play an important role in radial growth by increasing interfilament spacing; therefore they have direct influence on conduction velocity [38, 49, 65, 126]. Furthermore, as recent work has demonstrated, the Nfs are necessary for the development of dendritic arborization [134]. The Nfs can also affect the function of other cytoskeletal elements, such as microtubules and actin filaments, and may thus sway cellular motility [16, 111].

The Sternbergers produced and commercialized an antibody that gained wide acceptance and use, the SMI 32 [118], which reacts with a nonphosphorylated epitope in NfH of most mammalian species [46, 83]. The immunoreaction is masked when the epitope is phosphorylated. Curiously, phosphorylated and non-phosphorylated forms of Nfs can coexist in neurons [15]; 80% of axonal Nfs are phosphorylated, while 20% of them are not [81]. The immunoreaction against SMI 32 provides a characteristic Golgi-like staining pattern, which helps substantially to unveil previously unknown cytoarchitectonic boundaries or to confirm older ones. It specifically labels a subset of neurons, particularly dendrites, somata, initial axonal segments, and some rather thick axons widely across the central nervous system. In general, the Nfs are present in larger neurons with long and thicker axon [125] or with large and elaborate dendrites [79, 134]. Conversely, Nfs could not be found in the axonal growth cones or in dendritic spines, where actins and profilins play an important role in making these cellular structures capable of contracting and further movements [35, 99]. In addition, SMI 32 may serve as a useful marker in various neurodegenerative diseases, such as Alzheimer's disease (AD), Parkinson's disease (PD), amyotrophic lateral sclerosis (ALS), multiple sclerosis (MS), and diabetic neuropathy [87, 103]. Moreover, Nfs can be even be detected in non-neuronal tissue, e.g. in dermal Merkel cell carcinomas [98] (normally, the keratin is the characteristic intermediate filament of the epithelium [40]).

Petzold proposed [103] to name the immunoreaction product visualized with SMI 32 antibody to NfH^{SMI32}–ir (immunoreactive) element. Here, for the sake of simplicity, the same material it is simply labeled as SMI 32-ir element in the upcoming part of the chapter. There are several other commercialized or not commercialized antibodies that may also recognize Nfs. For instance the Lan3-8 [93], the Ta, Oc, Se [83], the rat-302 [58], the M14 and M20 [109], the N200 [110], the FNP-7 [67], the RabNF-M and RabNfH [74], the RMO255 and RMO55 [15] antibodies, and an array of SMI antibodies against the phosphorylated (SMI 31, 34, 35, 36, 310, 312) and the non-phosphorylated (SMI 32, 33, 37, 38, 39, 311) epitopes of Nfs [118].

This Nf proteins are specific to the mammalian brain; however, the similar protein triumviri of Nfs can be found in the avian [112] and Xenopus brain [120]. These non-mammalian isoforms' genetic background is akin to the mammalian ones and they may subserve analogous functions. It is likely that there was a common ancestor of vertebral Nf peptides and that precursor protein's sequence evolved in slightly different directions among species and preserved certain domains of the Nf ancestor [112, 83]. The other submammals also display comparable forms of proteins with analogous function. In crustaceans, no NfH and NfL can be detected; however, a NfM-like polypeptide is present in the crab *Ucides cordatus* [24]. Similarly, it was shown that only one Nf with high molecular weight (180kDa) is present in the fish brain forming functional polymers from this single type of protein [105]. Recently, another Nf with 50 kDa molecular weight has been cloned in lampreys, suggesting that some sort of assembly of Nfs can take place even in the agnathan nervous system [71].

Here in this chapter, I try to summarize our knowledge about the distribution of neurofilaments in the mammalian visual system by describing their appearance in the structures one by one from the retina to the cortex.

Retina

In general Nfs are used as neuronal markers in the central nervous system and also here in the retina. A group of large α-ganglion cells can be labeled with SMI 32 in the mouse [29, 30], in the cat [43], and also in the human retina [119]. Furthermore, the Nfs and neuropeptide Y (NPY) were colocalized in the large human retinal ganglion cells, raising the possibility that they are either different from the non-human ganglion cells or the non-human NPY-containing ganglion cells may also have Nfs [69, 119]. In addition, some amacrine and type A horizontal cells can be labeled with SMI 32-ir in the cat [43]. In another recent study, the Nfs were found to be present in the axons of horizontal cells, in the ganglion cells and in thin axons (probably belonging to amacrine cells) in the rabbit [130]. These findings support the notion of the presence of true axons in the retina. Most recently, it has been discovered that SMI 32 labels four different clusters of retinal ganglion cells in the mouse retina [23]. This grouping might point to a specific, but still unknown, function of these Nf-containing ganglion cells.

Similarly to what has been observed previously in non-mammalian vertebrates [136], NfH also accumulated in degenerating retinal fibers of rats. These fibers presumably innervated the dorsal lateral geniculate nucleus (dLGN) [29, 95], although the NfH accumulation was not seen in retinal fibers projecting to the superior colliculus [28]. In addition, the mRNA level of Nfs was also found to be increased during optic nerve regeneration in the goldfish [122]. This boost of Nfs in fibers may be due to the preserved thalamic contacts of the regenerating retinal axons [54].

A special carrying mechanism, the slow component a (SCa) of axonal transport, was first revealed and described in detail in the optic nerve of guinea pigs [8]. The SCa carries the Nf proteins and this transport depends upon the Nf proteins' phosphorylation state [115].

Concerning retinal pathology, human retinoblastoma cells and also horizontal cells in these eyes were devoid of Nfs [76]; however, a nuclear phosphoprotein isolated from human retinoblastomas showed homology with the NfLs [66]. This may indicate a defect in assembly of the functional Nf triplet. On the contrary, in animal ocular hypertension or glaucoma models, the amount of NfL was found to be significantly reduced [121]. Furthermore, in experimentally glaucomatous monkeys, the NfHs in ganglion cell axons were dephosphorylated, most likely indicating an Nf assembly deficiency in this disease also [73].

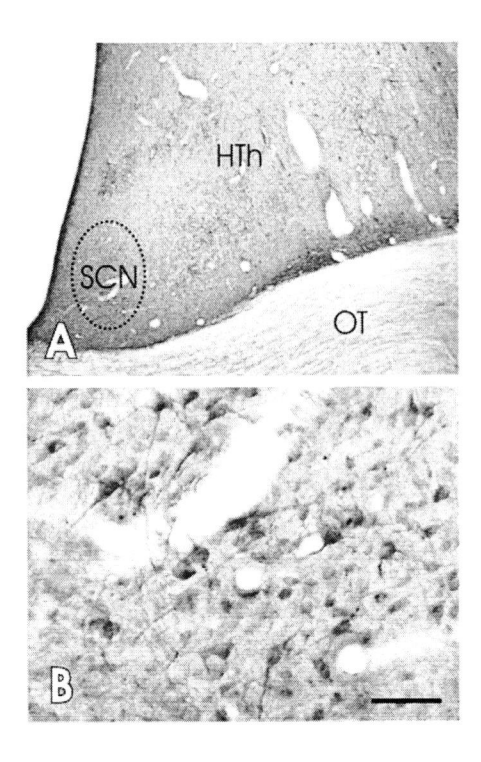

Figure 1. The feline suprachiasmatic nucleus is shown in low {A} and high {B} magnifications. SMI 32 immunolabeled perikarya and major dendrites can be seen {B}. Scale bar=500µm for A; 50µm for B.

Suprachiasmatic Nucleus

Although the hypothalamus of mammals in general is meager in Nfs, the suprachiasmatic nucleus (SCN) contains some of these proteins. To my knowledge, there is only one detailed study about the distribution of Nfs in the SCN [90]. This report describes an Nf-ir axonal meshwork in SCN forming clusters, a special neuronal organization which is characteristic of retinal axons elsewhere in the diencephalon. These clusters were found in both the ventral and the caudal part of the nucleus, and moreover these loci coincided with the known entrance areas of the retino-suprachiasmatic fibers. It has also been known that the NfH in the hypothalamus appears relatively late in development [36]. The reason for this might be the similar timing of hypothalamic maturation – which may also come out in a form of Nfs appearance – and the ingrowth of Nf-containing retinal fibers. It remains to be determined whether these Nf-ir axons really originate in the retina. On the other hand, cultured rat hypothalamic neurons from SCN were able to express phosphorylated Nfs, suggesting that neurons from the SCN under *in vivo* circumstances are capable of doing so as well [131]. This raises the question of whether the observed Nf-containing axons are really of extrinsic origin (i.e., retinal) or whether it is an intrinsic property of SCN neurons.

In our material, perikaryal and neuropil labeling of SMI 32 were both detected in the rodent and in the feline SCN (Figure 1.). These cells were about 10-20µm in diameter and showed both fusiform and multipolar characteristics. Their well-stained dendrites could be followed for as long as 100μm.

Brainstem Oculomotor Nuclei

In general, ventral horn motor neurons of the spinal cord [46, 63, 64, 123] and also those in the brainstem motor nuclei [55, 82, 123] are abundant in Nfs. They also showed immunoreactivity for the SMI 32 antibody [20, 47].

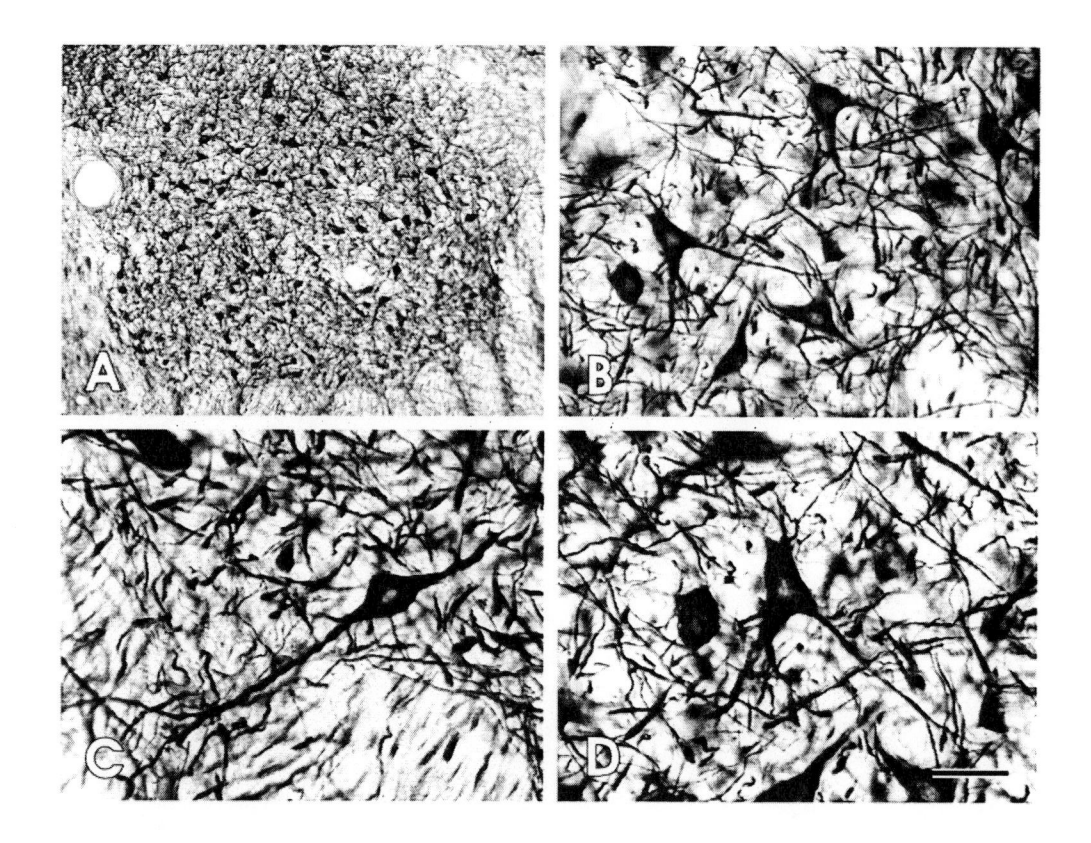

Figure 2. Nucleus of the oculomotor nerve located in the rostral and tegmental part of midbrain of the macaque {A}. The large oculomotor neurons are intensely stained for SMI 32-ir {B-D}. Scale bar=500μm for A; 50μm for B, 33μm for C, D.

In our material, large ocular motor neurons were readily detected in the brainstem of rat, cat and monkey, which adequately delineated the position of the oculomotor nucleus (nIII; Figure 2.). The trochlear and abducens nuclei were also easily found in the mesencephalic and pontin tegmenta of these species. These motoneurons were fairly large (20-30μm in diameter), possessed thick dendrites that branched relatively rarely, and could be followed for a long distance. In oculomotor nuclei, a strong neuropil label was observed consisting of strongly-stained SMI 32-ir dendrites. This striking feature enabled an extraordinarily precise histological demarcation of the nuclei, one of the best in the brain. The distribution and appearance of Nfs in these nuclei showed virtually no interspecies differences among rodents, carnivores or primates, which may substantiate the relatively conserved neuronal organization of the oculomotor system.

Furthermore, the Nfs allowed a cytochemical distinction between neurons in the nIII innervating single or multiple extraocular muscle fibers, both in frontal eyed and lateral eyed species [33]. Those neurons innervating single muscle fibers were localized within the classical cytoarchitectural borders of oculomotor, trochlear and abducens nuclei and stained for SMI 32, while those innervating multiple muscle fibers in nIII did not. These cells were located at or around the histological boundaries of the nIII [33].

At the cellular level, large motor neurons require the NfL subunit for dendritic growth and for building elaborate dendrites [79, 134]. On the other hand, a decline of NfM content with age has been specifically detected in cat sensory and motor brainstem structures, such as in facial motor, gracile and cuneate nuclei [133]. It may be possible that the same phenomenon can also be observed in the brainstem oculomotor nuclei.

Superior Colliculus

Thorough descriptions of SMI 32-ir neuronal elements are available in the cat [42] and in the marmoset [12]. SMI 32-immunohistochemistry provided a Golgi-like cellular labeling pattern in both species, since it was present in somata, and the stain could be followed almost till the tertiary dendrites. The staining was the strongest in stratum griseum superficiale (SGS), whereas it resulted in moderate neuropil label in stratum opticum (SO) and in stratum griseum intermedium (SGI; see also Figure 4.). The uppermost stratum zonale (SZ) was meager in label (with the exception of some dendrites from the layer beneath). The SGS contained some smaller neurons with various morphology. In the stratum griseum profundum

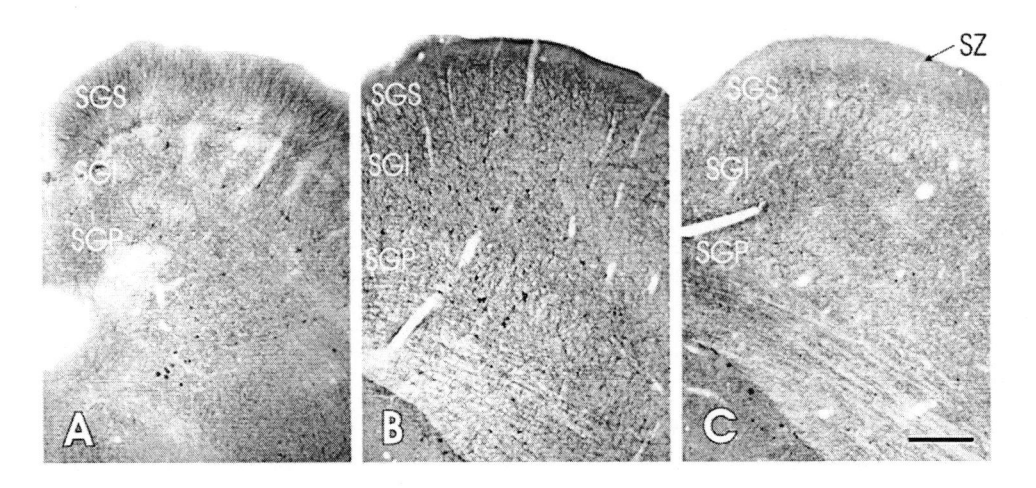

Figure 3. Low power photomicrographs showing the distribution of SMI 32-ir neuronal element in the rodent {A}, feline {B} and macaque {C} superior colliculi. Whereas the uppermost layer I, and the SZ is weakly labeled and the three grey layers (SGS, SGI, SGP) show heavier SMI 32-ir. Scale bar=500μm.

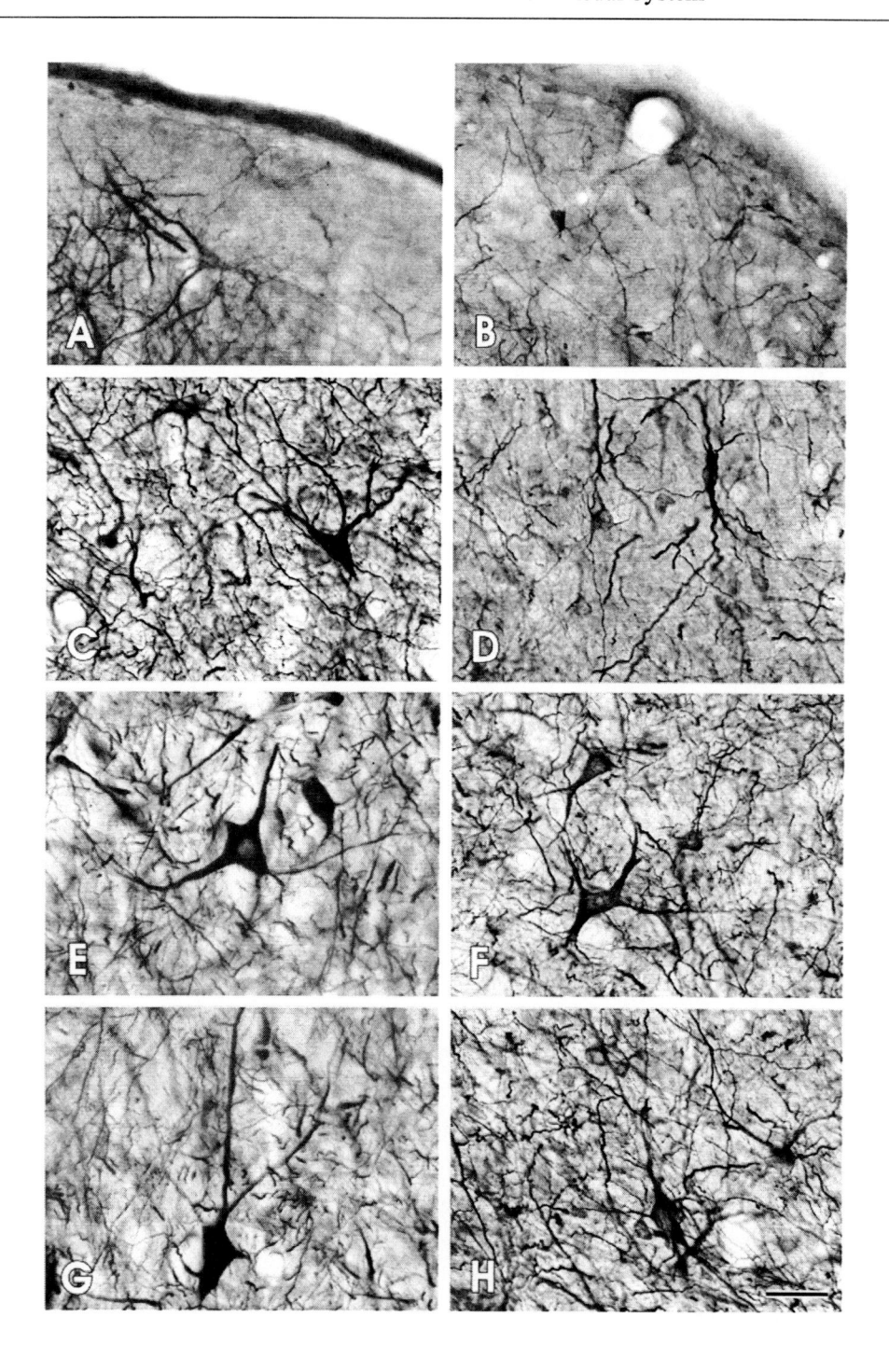

Figure 4. High power photomicrographs illustrate SMI 32-ir elements in different layers of the superior colliculus. The SZ and the SGS are shown in the cat {A, C} and in the macaque {B, D}. Large multipolar neurons can be seen in the SGI of the cat {E} and the macaque {F}. Characteristic large fusiform neurons are found in the SGP of the cat {G} and of the macaque {H}. Scale bar=50μm for A, B, E, G; 33μm for C, D, F, H.

(SGP) the SMI 32 stained patches which contained some very conspicuous and intensely labeled large neurons. These quite characteristic, probably motor-related output cells, showed remarkable similarity to the motoneurons of the oculomotor nuclei [92]. It often showed in a pyramidal form with an apical dendrite orienting toward the collicular surface, and rarely other appearances occurred too. The Nf-containing neuronal elements formed patches and established column-like interconnections with the similarly organized deeper layers. The pyramids in the deep layers were also intensely immunostained for SMI 32 in these reports (see also Figure 4.). Recently, these Nf-ir SGP pyramidal neurons all proved to be brainstem projecting neurons innervating motor structures in the cat, and they also contained parvalbumin [42].

The rat, cat and macaque superior colliculus (SC) appeared to have very similar distribution of SMI 32-ir neuronal elements in our material (Figure 3.). The SMI 32 staining pattern was overtly akin to the previous results described above, even in rodents. The Old World monkey macaque has also showed verisimilitude to the New World monkey marmoset. The characteristic tectal feature, the columnar association of SMI 32-ir elements in intermediate and deeper layers, could also be observed in all these species, advocating a universal organizational principle of the SC.

In addition, some $GABA_C$ receptor containing cells exhibited Nf-ir in the rat collicular superficial layers [22]. There is no other study available about the occurrence of Nfs in the rodent superior colliculus; only an atlas provided low magnification images of SMI 32 staining in the rat midbrain [101]. The occurrence of Nfs in the deep layer pyramids of the SC seems to be rather ubiquitous among species, since it was even present in the fish [89], in the frog [108] and in the chicken tectum [127]. As discussed above in this chapter, these non-mammalian isoforms of Nfs are dissimilar immunogenetically and are also termed differently. In fact, the observed wide presence of functionally similar Nfs in the deep layer pyramids of superior colliculi or tecta of the animal kingdom may further imply an ancient feature of the organization of the oculomotor system.

Accessory Optic System

No study of the distribution of Nfs in this ancient, but tiny and therefore often neglected system is available [45]. In our material, the immunohistochemical appearance of SMI 32 label in the feline terminal nuclei (dorsal, lateral and medial) of the accessory optic system (AOS) was found to be very similar to the pretectal nucleus of the optic tract (NOT; Fig 5.). Nonetheless, the intensity of SMI 32 staining within the AOS was less strong than what could be seen in the collicular SGS. In general, the SMI 32-ir elements allowed a fine histological identification of these minuscule nuclei. Some multipolar and fusiform perikarya have been labeled with well-stained dendrites resulting in a medium neuropil label in the AOS. The appearance and distribution of SMI 32 neuronal elements were almost identical among the rat, cat and macaque. Likewise, the distribution of Nf containing elements was very similar throughout the accessory optic nuclei; however, in the dorsal terminal nucleus more SMI 32-ir perikarya could be observed. The lack of interspecies and internuclear variability in Nf distribution proved further that the AOS is one of the oldest, perhaps the most conserved part of the mammalian visual system.

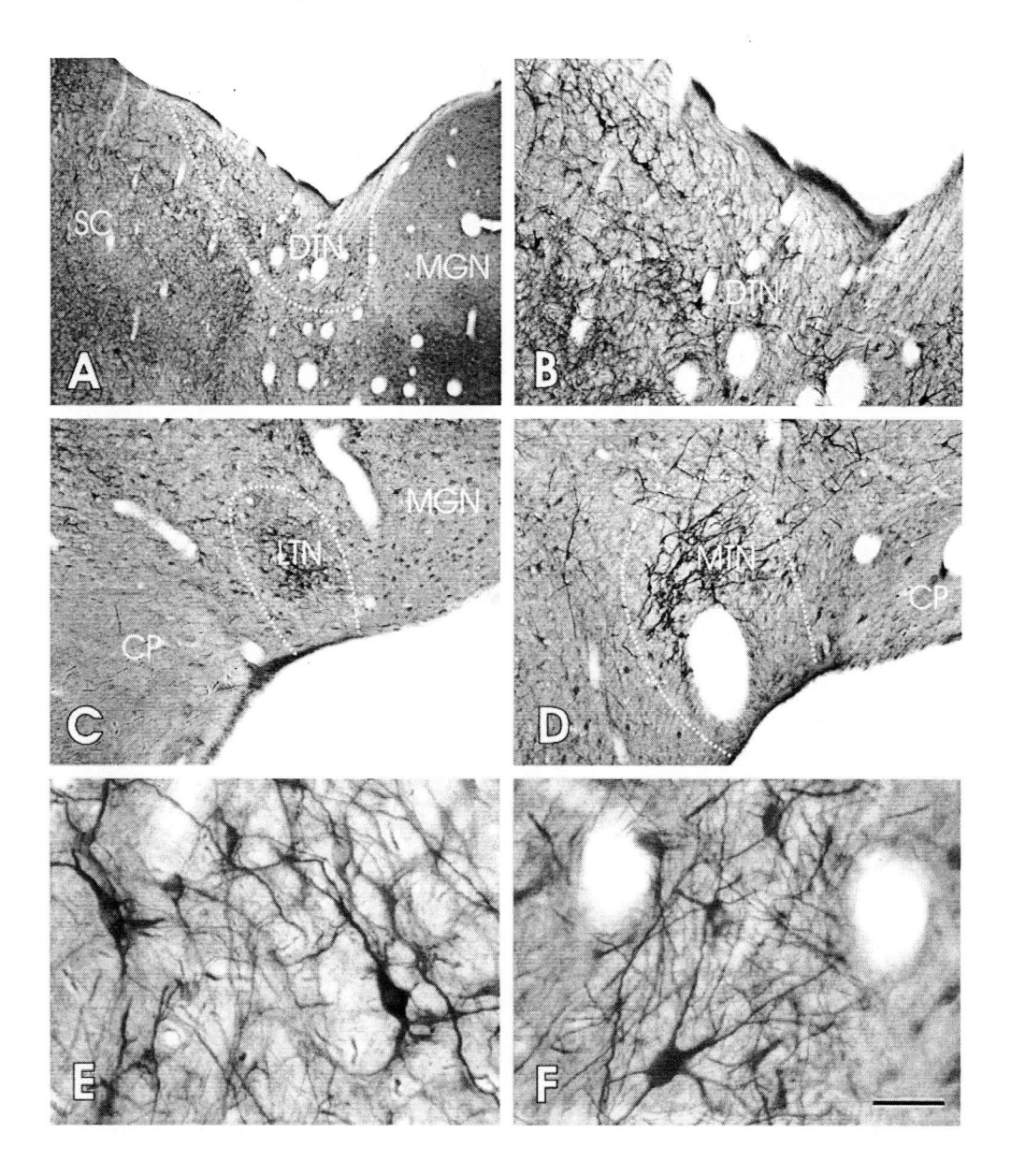

Figure 5. The presence of Nfs can be detected in all three terminal nuclei of the accessory optic system: in the DTN {A, B}, in the LTN {C} and in the MTN {D}. The appearance of SMI 32-ir neurons was similar to those in the NOT {E, F}. Scale bar = 500μm for A; 200μm for B-D; 50μm for E, F.

Pretectum

In the pretectum, the retinal axons running toward the thalamus are still glia-ensheathed and greatly intermingled with intrinsic neurons. This is particularly true in the NOT, as its name also suggests. In fact, glial elements are copiously present in the entire pretectal nuclear complex [44, 132]. Unfortunately, only a few data are available on the occurrence of Nfs in the pretectum [5, 94]. As these works reported, the neuronal elements of the pretectum were rich in Nfs, though in general the intensity of SMI 32 immunolabeling was less intense than that of the SC and stronger than the thalamic middling. In our experience, SMI 32 labeled pretectal cells throughout the nuclear compound with various morphology (multipolar, fusiform) in all species investigated so far. The dendrites could best be visualized with SMI 32 and could be traced a long way from the somata.

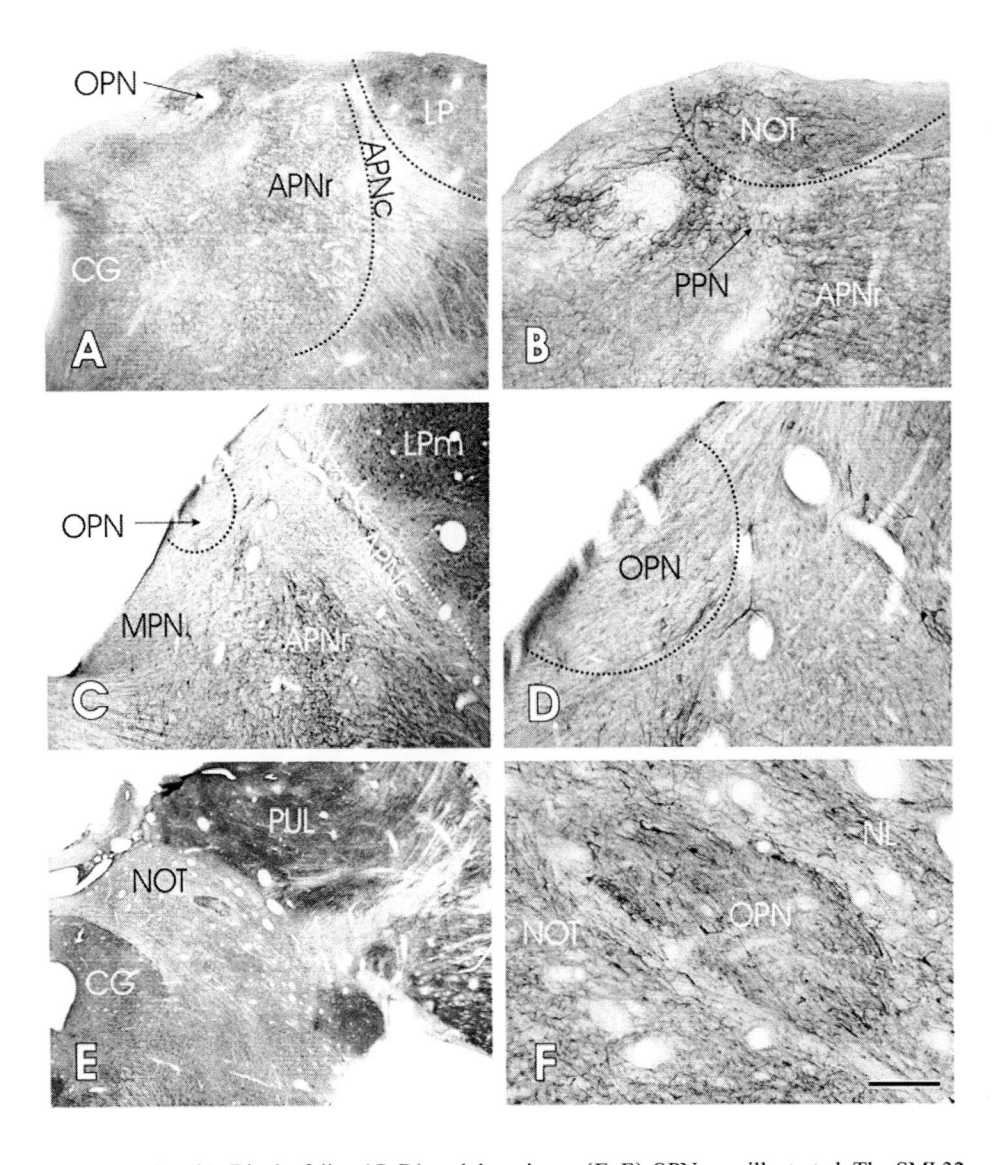

Figure 6. The rodent {A, B}, the feline {C, D} and the primate {E, F} OPNs are illustrated. The SMI 32 staining delineates the OPN particularly well in the rat and the macaque. The rodent NOT and PPN can also be easily seen {B}. The APNc labels the pretecto-thalamic border in all three species {A, C, E} and in the macaque even the nucleus limitans can be observed {F}. Scale bar = 500µm for A, C; 200µm for B, D, F; 1600µm for E.

The best known structure of the nuclear complex is perhaps the olivary pretectal nucleus (OPN) [44, 77]. The SMI 32-ir neurons lying at its border were almost wrapping the structure, while the inner nucleus was meager in Nfs in the rat and cat but not in the monkey (Figure 6.). These SMI 32-ir neurons were fairly large (15-25µm in diameter), but some smaller (10-15µm) cells could also be observed and they showed various morphology. However, the dominant cell shape of the OPN was still fusiform at its nuclear border (Figure 7.). The NOT, on the other hand, showed stronger SMI 32-ir neuropil labeling than the OPN. The Nf-containing cells of the NOT were of similar size to the OPN's and also presented various morphology (Figure 8.). Similarly, as observed previously for NPY-ir neuronal elements of the feline NOT [10], the SMI 32-ir NOT cells also displayed two sets of forms (a fusiform and a multipolar) in rodents, feline and primate pretecta [5].

In addition, slight cytoarchitectonic delineation was possible between the NOT and the posterior pretectal nucleus (PPN), where the dual morphology of SMI 32-ir neurons was also present (Figure 6B). Unfortunately, the SMI 32 could not break through the obstacles of cytoarchitectonic identification of the medial pretectal nucleus (MPN).

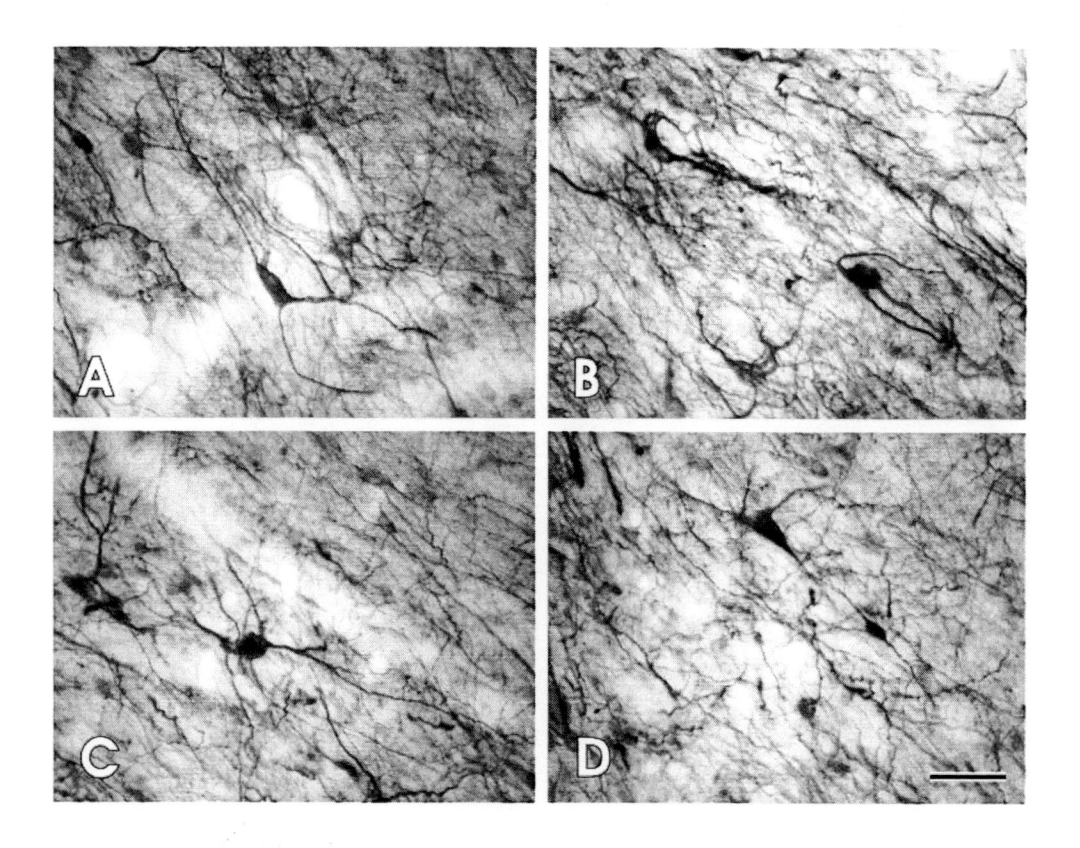

Figure 7. Neuronal elements stained with SMI 32 in the monkey OPN. The cells are mostly fusiform, but multipolar neurons also found. The SMI 32-ir neurons are localized at the cytoarchitectonic border of the OPN {A-D}. Scale bar = 50µm for A-D.

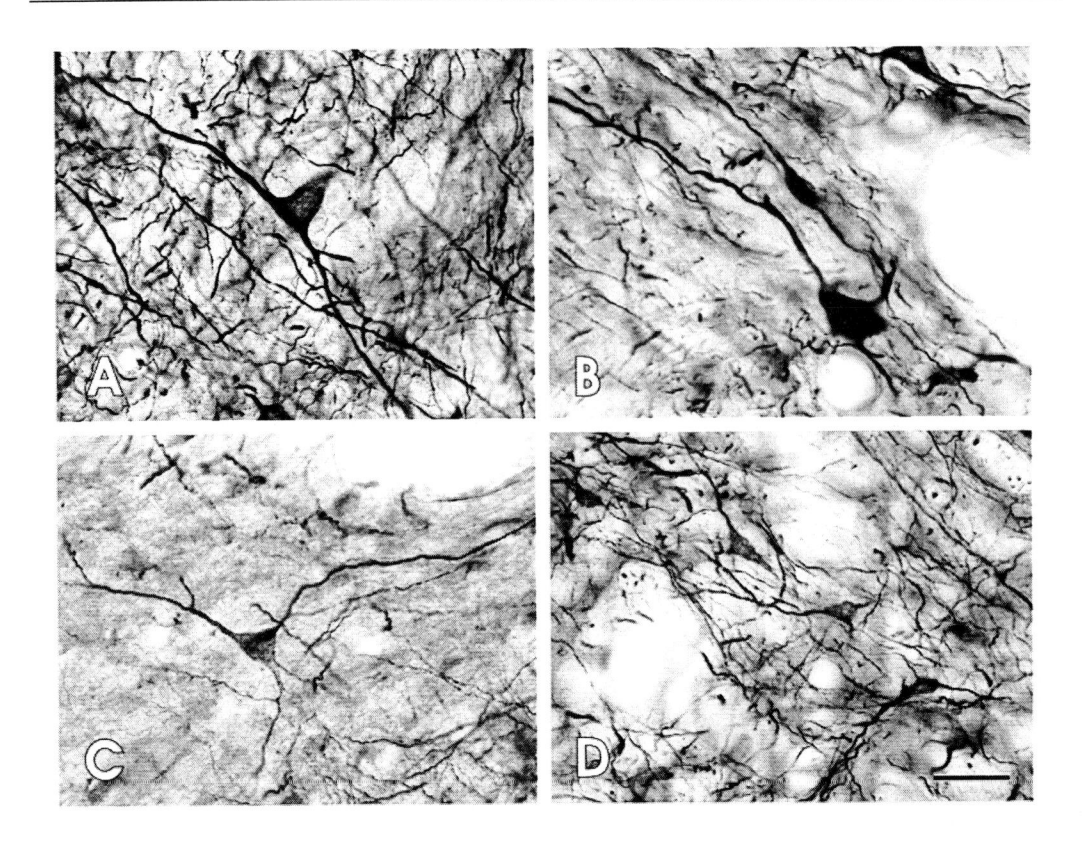

Figure 8. Neuronal elements stained with SMI 32 in the monkey NOT. The cells are mostly multipolar, but some fusiform neurons also found. The dendrites can be followed for a long way from the soma. Scale bar = 33μm for A-D.

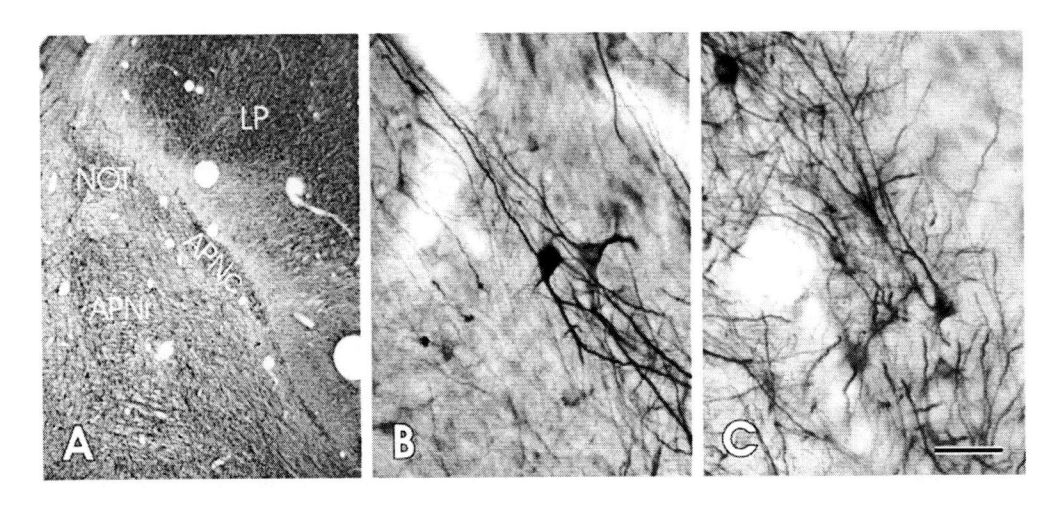

Figure 9. The two portions of the anterior pretectal nucleus are easily differentiated in the rostral part of the cat pretectum {A}. The compact portion borders the thalamus and contains fusiform neurons {B, C}, while the reticular portion composes the mass of the rostral pretectum {A}. The fore tip of the NOT can still be noticed {A}. Scale bar = 500μm for A; 50μm for B, C.

In the compact portion of the mammalian anterior pretectal nucleus (APNc) the SMI 32-ir neurons were mostly fusiform and localized at the pretecto-thalamic border (Figure 6.). This feature was also described in the case of NPY-ir cells of the feline APN [10]. Their easily-detectable dendrites ran in parallel with the pretecto-thalamic cytoarchitectonic border. This subnucleus was abundantly stained with SMI 32 compared to the somewhat concealed reticular portion (APNr). In APNr, the SMI 32-ir cells were less dense and predominantly multipolar; likewise in rats, cats and primates. In general, the pretectal SMI 32-ir cells and their neuronal organization across the nuclear complex closely resembled the pretectal NPY system, raising the possibility that they are perhaps identical. Additionally, the SMI 32 immunolabeling permitted the chemoanatomical identification of another little-known structure, the nucleus limitans that is also localized at this intradiencephalic pretecto-thalamic border of embryologic origin (Figures 6, 9A).

Visual Thalamus

Dorsal Lateral Geniculate Nucleus (dLGN)

In monkeys, the neurons of the geniculate magnocellular layers (principal thalamic relay station of the magno- or M-stream, chiefly analyzing motion) stained stronger with SMI 32 than the parvocellular or koniocellular layers both in Old World [21, 52, 106] and in New World monkeys [11]. Similarly, in our primate material, the magnocellular neurons of dLGN stained the strongest, the parvocellular lighter, while the koniocellular layers almost lacked Nf-ir (Figure 9B; see also Figure 2 of [11]). We could observe only SMI 32-ir relay cells in the dLGN, but no small interneurons in any species investigated. The diameter of labeled neurons in macaque magnocellular laminae was found to be about 15-25 μm, whereas that of the parvocellular cells was observed to be considerably smaller (approximately 10-15 μm). In experimentally glaucomatous macaques, the Nf immunolabeling was selectively reduced in geniculate parvocellular layers [128].

In cats, the SMI 32 proved to be a true marker for geniculate cells of the Y visual stream (homologue of the primate M-pathway) [7]. In the feline geniculate laminae A, A1, C_{magno} and in the medial intralaminar nucleus (MIN), large Y-cells were well stained with SMI 32 (Figures 11A, B), whereas in C_{parvo} layers only a light label was present (Figure 10D). This staining pattern corresponded well with the functionally homologous layers of the primate LGN. Following monocular lid suture of cats, the staining of SMI 32-ir neurons were reduced in the deprived geniculate laminae, indicating a cytoskeletal change [7]. This structural alteration was attributable rather to cell class competition and not binocular competition in the feline dLGN. Thus, the nature of retinal input that the geniculate Y-cells receive may have a major influence on the organization of the neurons' Nfs. The presence of Nfs in these neurons in turn may add some advantages to the cells' success. Ultrastructurally, Nfs and microtubules were seen in close apposition of the postsynaptic membrane of asymmetrical synaptic contacts in geniculate cells [72].

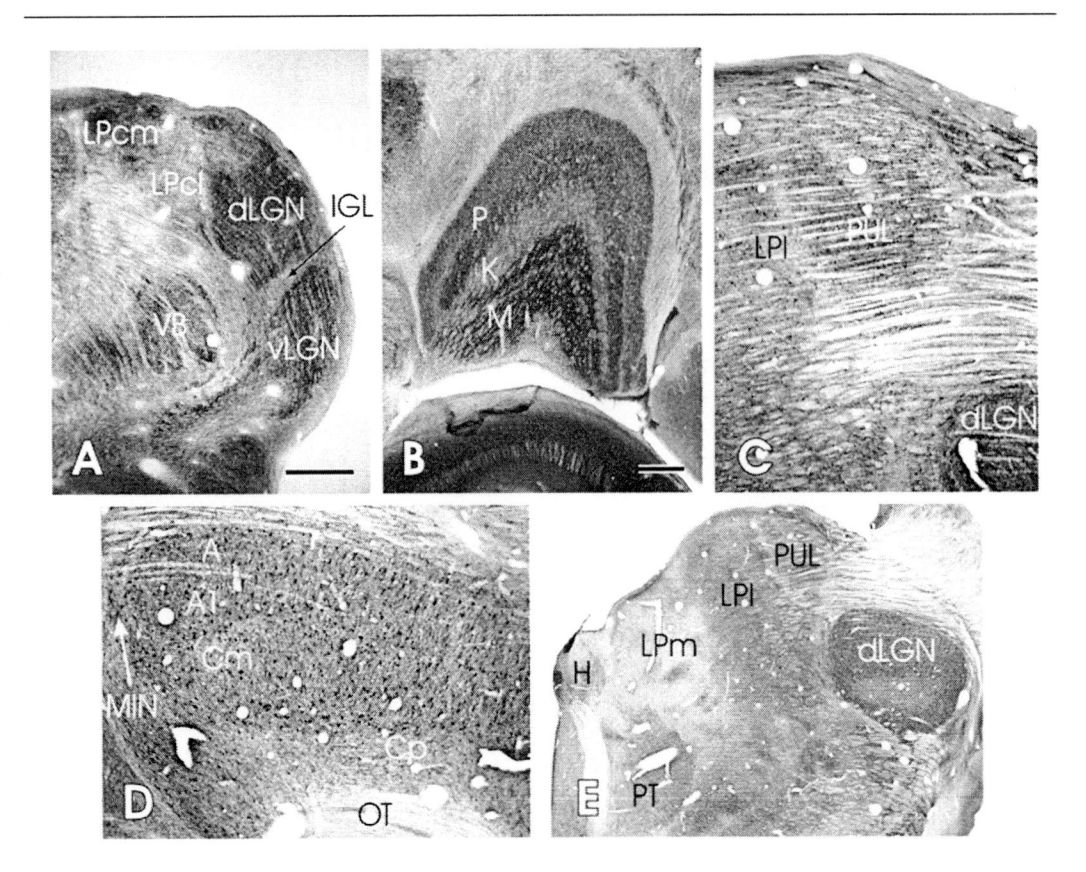

Figure 10. The dorsal lateral geniculate nuclei in the rat {A}, in the cat {D} and in a New World monkey, the Callithrix jacchus {B} are shown. The lamination in the monkey and cat dLGN is also illustrated. The SMI 32 distinguishes the parvo-, konio-, magnocellular layers of monkey dLGN {B}, and makes nuclear parcellation possible in all species {A, E}. The otherwise veiled cytoarchitectonic boundaries of the feline pulvinar can be easily noticed {C, E} and the SMI 32 staining permits the differentiation of the medial and lateral portions of the feline LP {E}. In addition, it demarcates the thalamus from the pretectum {E}. In the rodent LP, the medial and the lateral subdivisions can be seen {A}. Scale bar on A = 500µm for A, 200µm for B, C; 1600µm for E. Scale bar on B = 500µm.

Since the SMI 32 was merely capable of distinguishing relay cells from interneurons in rodents [70], it probably cannot serve as a morphological basis for cell class differentiation. However, it perfectly delineated the nucleus [102].

In monotremata, the SMI 32 labeled some larger neurons in the LGb (dLGN homologue), supporting the notion that the presence of Nfs is a true marker for the magnocellular visual pathway even in such an ancient species [3].

Following enucleation of macaques, the degeneration of retinal terminals could be seen and neurofilamentous hyperplasia forming annular bundles (or synaptic rings) were found ultrastructurally in these boutons [100]. This electron microscopic feature has also been called neurofilamentous or dark degeneration in observations made in night-active New World monkeys [53]. In another study, increased NfM expression in the degenerating retinal nerve endings was found in enucleated rats [19]. More recently, the degeneration of dLGN neurons following rat striate cortical lesion was characterized by aberrant accumulation of perikaryal non-phosphorylated Nfs and by early vacuolation and subsequent swelling of dendrites at the ultrastructural level [1].

There is another marker that specifically labels the magnocellular or Y-cells in the LGN, the cat-301 [59]. This monoclonal antibody recognizes extracellular surface proteoglycans composing the perineuronal net, while the SMI 32 is a faithful marker of the intracellular Nf cytoskeleton. Nonetheless, there is no comparative study available comparing the two substances in the thalamus or detecting possible colocalization.

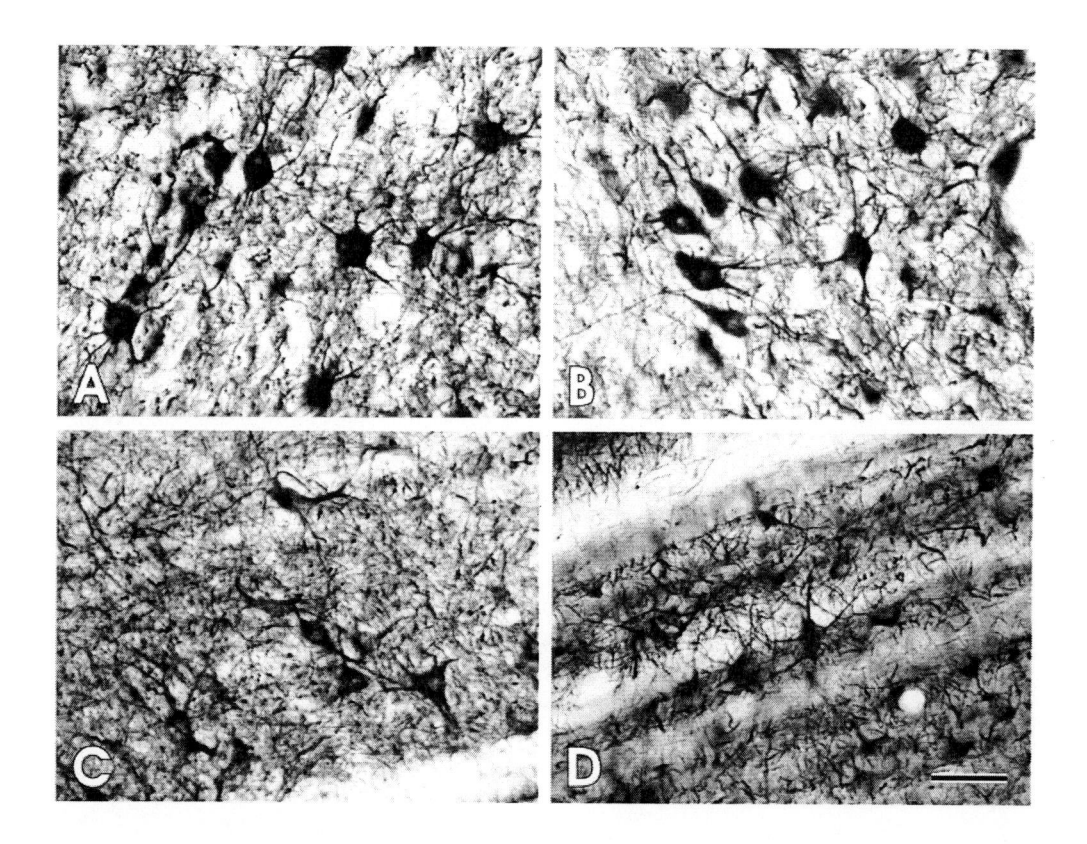

Figure 11. High power photographs illustrate the cat thalamic SMI 32-ir neurons of multipolar morphology. Large relay cells of the dLGN lamina A are shown {A, B}, whereas in the LP {C} and the pulvinar {D} smaller cells can be noticed. Scale bar = 50μm.

Pulvinar (PUL)

In the monkey, the SMI 32 immunostaining resulted in an overall light staining pattern (especially compared to the neighboring LGN), with some scattered perikaryal and proximal dendritic label. Although relatively few neurons were labeled in the pulvinar as a whole, some well-stained neurons were observed both in Old World [52] and in New World [116] monkeys. The label helped to chemoarchitectonically parcellate the large thalamic nuclear complex of monkeys. The medial division of the inferior pulvinar (PI$_M$) was distinguished by its numerous SMI 32-ir neurons. The small posterior division (PI$_P$) did not present immunoreactive cells for Nfs, making it easily recognizable [52]. The distribution of SMI 32-ir elements in the pulvinar of the New World monkey, *Cebus apella,* also resulted in similar subnuclear parcellation [116].

In our material in the cat homologous nucleus - the lateral posterior pulvinar complex (LPP) - the SMI 32 staining was observed to be largely light, resulting in thalamic relay neuron label and no interneurons. However, the pulvinar 'proper' could be neatly delineated (Figure 10C), and the medial and lateral divisions of the LP could be differentiated (LPm, LPl; Figure 10E). While the acetyl cholinesterase reaction as a cytochemical means of the intranuclear differentiation of LPP has long been known [48], the SMI 32 staining proved to be a second useful histological marker enabling the same distinction. Whereas the label was stronger in pulvinar 'proper' and in LPl, the LPm presented a weaker stain. The SMI 32-ir neurons were about the same size in the LPl (Figure 11C) and in the pulvinar 'proper' (Figure 11D). While LPP neurons in general possessed similarly frequently branching and elaborate dendrites as the geniculate Y-cells, they were a bit smaller in size and less intensely stained.

Unfortunately there is no detailed study available of the rodent lateral posterior nucleus (LP), only a low magnification figure of SMI 32 staining in LP is shown in a cytoarchitectonic atlas [102]. In our material, the SMI 32-ir cells in LP could not faithfully delineate the rostral and the caudal subdivisions; however, it could adequately parcellate the caudal LP into medial (LPcm) and lateral (LPcl) subdivisions (Figure 10A). The LPcl presented weaker SMI 32 label, whereas the LPcm stained more intensely. The subnuclei were well-demarcated from the medially placed pretectal mass and the laterally lying dLGN.

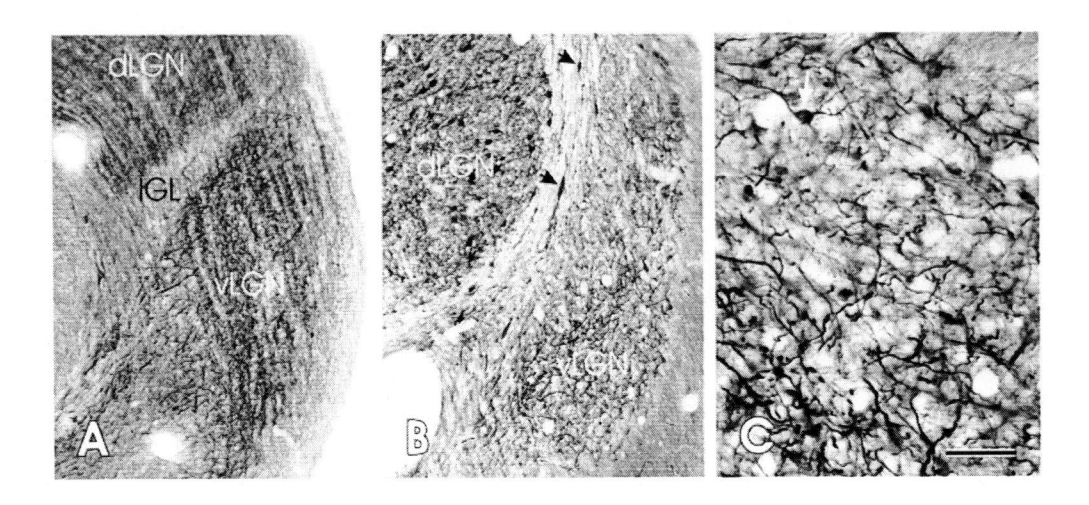

Figure 12. The rodent {A} and feline {B} ventral lateral geniculate nucleus is shown. While a strong SMI 32-ir neuropil is present in the dorsal thalamus, the vLGN shows a looser neuropil labeling, especially in the cat {B}. The rodent IGL is almost devoid of labeling {A}. Some typically fusiform cells can be observed in the TRN between the dLGN and the vLGN {black arrowheads; B}. On high power images strong dendritic label and several petite SMI 32-ir cells can be seen in the vLGN {white arrow; C}. Scale bar = 200μm for A, B; 50μm for C.

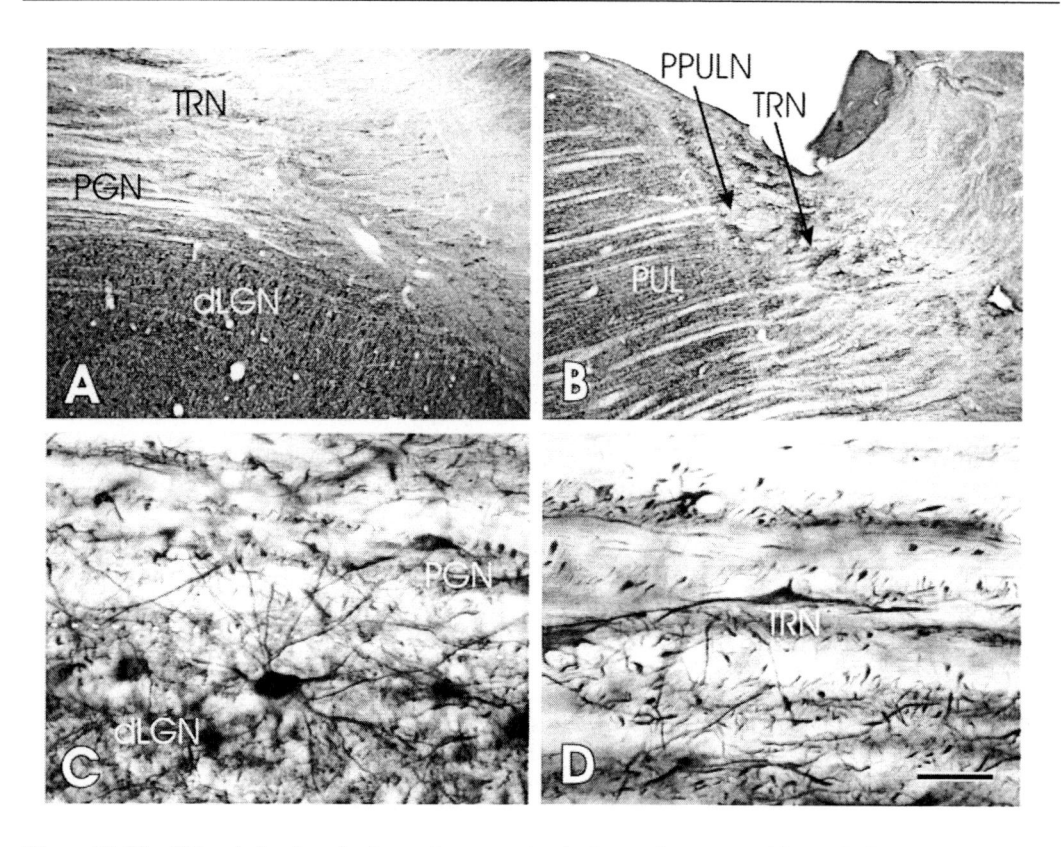

Figure 13. The feline thalamic reticular nuclear complex is shown. Its outer and inner shells wrapping the dLGN (PGN) and the pulvinar (PPULN) are illustrated {A, B}. An SMI 32-ir geniculate relay cells infiltrates the PGN with its long dendrites {C}, and a typical reticular fusiform neuron spanning its dendrites along the TRN can be seen {D}. Scale bar = 500μm for A, B; 50μm for C, D.

Recently, the presence of Nfs in thalamic relay cells has also been detected in the echidna [3]. Similarly to the eutherian species above, the SMI 32 staining resulted in a lighter label in the echidna LP. While a poor neuropil label could be observed, some larger perikarya showed Nf immunoreactivity in the LP.

Thalamic Reticular Nucleus (TRN)

In all investigated species, the neurons of the TRN presented intense SMI 32 stain (Figure 13D). In our material, these cells showed typical fusiform morphology with flattened dendrites in the plane of the nuclear mantle and were rather scattered along the TRN (Figures 13A, B). In addition, SMI 32-ir effectively visualized the two shells of the feline TRN 'complex'; the outer TRN 'proper' and the inner perigeniculate and the peripulvinar nuclei (PGN, PPULN; Figure 13B). The TRN 'proper' contained somewhat more neurons with intense SMI 32-ir and the cells were all fusiform (Figure 13D), whereas the PGN cells were less immunoreactive, less in number and rather rounded (Figure 13C). As the SMI 32 stain demonstrated chemoanatomically, the outer shell (TRN 'proper') of the reticular nucleus was observed to be continuous with the ventral division of the zona incerta, while the inner shell

(PGN/PPULN) with the dorsal division of this ventral thalamic nucleus. Curiously, this bilamination of the TRN 'complex' was not obvious in rodents and macaques, but was present in the New World monkey marmoset [3]. It was also observed in the cat that the SMI 32-ir dendrites of lamina A relay cells often span into the PGN, but not into the TRN 'proper' (Figure 13C). This finding further confirmed the presence of geniculate relay cell dendrites in the cat PGN, which was observed previously at the ultrastructural level [26]. Moreover, it has recently been described that the SMI 311 (another antibody against phosphorylated Nfs) stained the developing human TRN [124].

Ultrastructurally, bundles of Nfs run along the plasma membranes of TRN dendrites [85], which are able to form non-synaptic symmetrical filamentous contacts with one another. These connections may mediate intercellular non-synaptic communication amongst TRN cells, probably involving gap junctions. This type of link ensures rapid spreads of excitation within the TRN, makes possible rhythm generation and pacemaker activity, which features have long been described in this nucleus [117]. The presence of such connections may explain the abundant Nf content of the reticular neurons.

Suprageniculate Nucleus (SGN)

There is no study available reporting on the distribution of Nfs in SGN. In our material, the distribution of SMI 32-ir elements in the SGN was very similar to the 'associative' or 'higher order' thalamic lateral posterior nucleus and its homologues in rodents, carnivores and primates. The immunostained SGN cells were about the same size as the LP relay cells and also the dendritic branching pattern was very much alike in all investigated species.

Visual Cortex

Cortical SMI 32 immunostaining was employed first in the occipital lobe of normal macaques and humans [17]. Shortly thereafter, the pathological research began with the examination of the cortex of patients with Huntington's disease [25, 88]. Since then, the Nf staining of brain tissue – and also the detection of Nfs in cerebrospinal fluid indicating axon loss – has gradually gained larger medical importance [103, 108].

One of the original reasons for the use of SMI 32 was to be able to parcellate otherwise homogenously stained cortical areas with classical means (e.g. Nissl stain, see also Figure 14). Fortunately, up to now a long series of studies described the areal distribution of SMI 32 in various cortical regions of a wide collection of species: echidna, *Tachyglossus aculeatus* [56], wallaby, *Macropus eugenii* [4], whale, *Megaptera novaengliae* [60], mouse, *Mus musculus* [50], rat, *Rattus norvegicus* [75], hamster, *Mesocricetus auratus* [9], cat, *Felis catus* [51], dog, *Canis familiaris* [62], marmoset, *Callithrix jacchus* [6], vervet monkey, *Cercopithecus aethiops* [21], cynomolgus monkey, *Macaca fascicularis* [61] and human, *Homo sapiens* [17, 107]. While many maps have been set out for the cortex, few of them are complete. From February 2006, the distribution of SMI 32-ir elements in the entire macaque's hemisphere can also be viewed online (http://brainmaps.org/index.php?action

=viewslides&datid=15) at a very high magnification on superb quality images. In addition, complete atlases with SMI 32 stain for the rodent brain are also commercially available [101, 102].

In general, SMI 32 is present in the somatodendritic compartment of pyramidal cells situated in supra- and infra-granular layers [17] of the cortex. A large pool of Nf-containing neurons can be found in layer III and V (Figure 15A), and to a lesser extent in layer II and VI, which show some areal variability (Figure 14). Specifically, among infragranular cells the large subcortically projecting pyramidal neurons (the 'type I' layer V projection neuron, traditionally named Betz cell) are SMI 32-ir stained, while callosal projecting neurons (the 'type II' layer V projection neuron) are not [18, 96, 129]. In the cortex, however, another subset of pyramidal neurons - presumably cortico-cortical ones - are stained for SMI 32 in layer III. In layer V, these pyramidal neurons grouped together (Figure 14C), whereas in layer III they are distributed rather evenly. Nevertheless, some non-pyramidal neurons were also found to be SMI 32-ir stained in the supragranular layers (Figure 14D-F): a fusiform cell (Figure 14D), a multipolar cell (Figure 14E) and another cell type closely resembling spiny stellate neurons (Figure 14F).

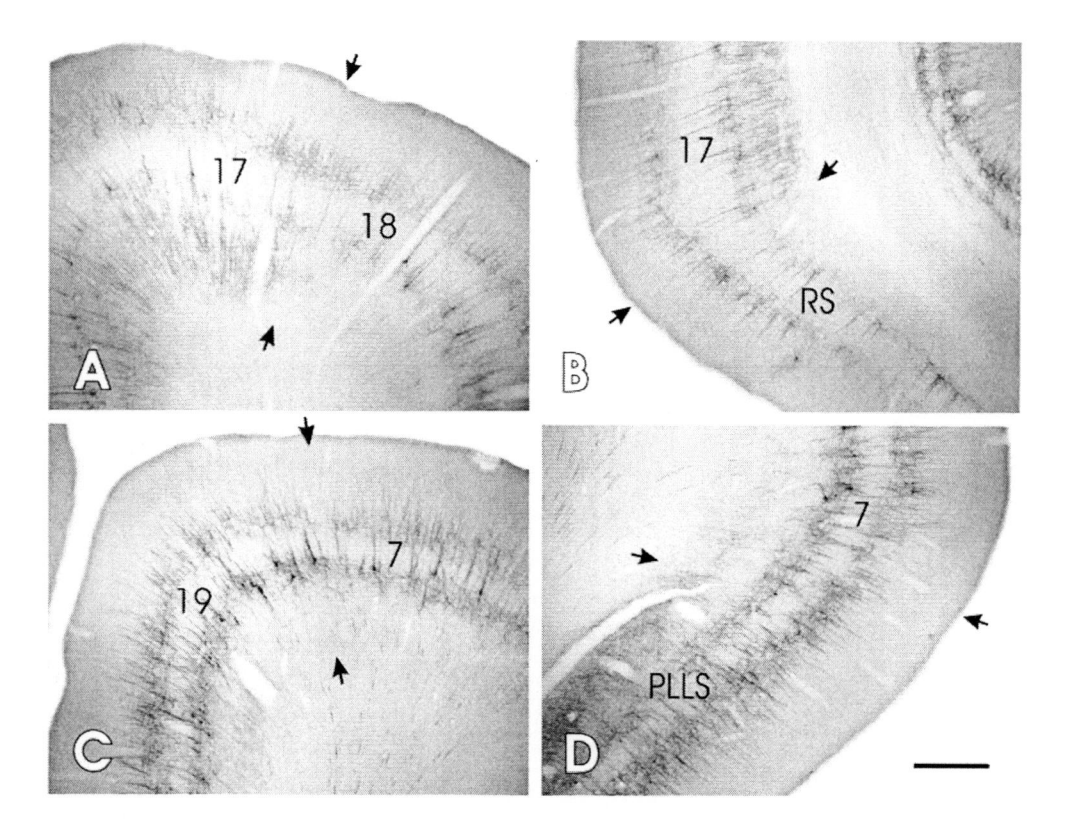

Figure 14. Cytoarchitectonic areal boundaries revealed by SMI 32 stain are illustrated (arrows). The different distribution of supragranular and infragranular neurons serves as the basis of cortical parcellation. Borders between area 17 and area 18 {A}, area 17 and retrosplenial cortex {B}, area 19 and area 7 {C}, area 7 and posterolateral lateral suprasylvian cortex {D} are shown. Scale bar = 50μm.

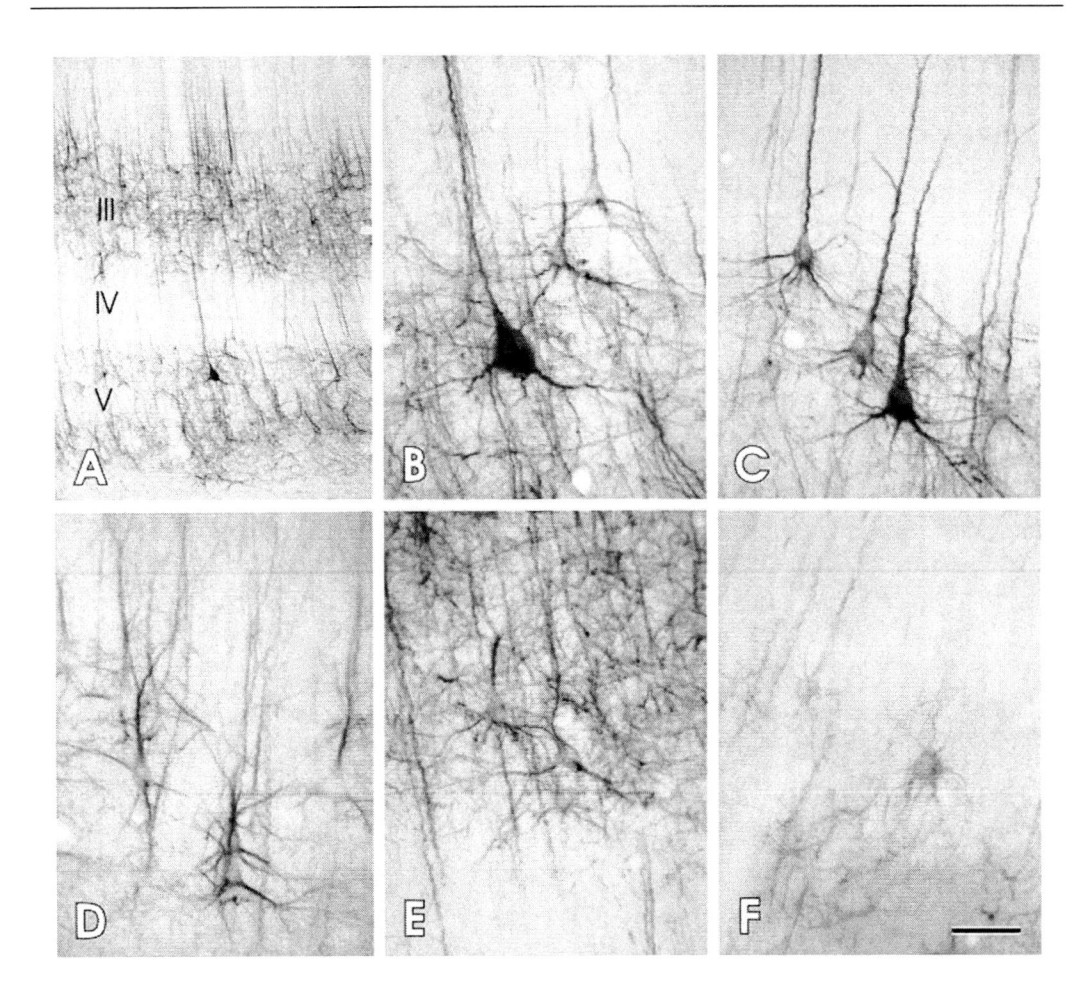

Figure 15. SMI 32-ir neurons show bilaminar distribution in the cat area 17 {A}. High power photographs illustrate the SMI 32-ir neurons in the infra- {B-C} and supragranular layers {D-F}. A large layer V pyramid can be seen {B} and further SMI 32-ir neurons group close together {C}. The pyramids in layer III distribute more evenly {D, E} and sometimes SMI 32-ir neurons resembling spiny stellate can be observed {F}. Scale bar = 200µm for A; 50µm for B-F.

The SMI 32 could visualize various boundaries within the visual cortical orchestra. It was first described in the rhesus monkey [61], then in the vervet monkey [21], in the cat [51], and most recently in the marmoset [6]. The bedrock similarities - serving areal parcellation purposes - between Old World and New World monkeys' SMI-32-ir were described as: (i) in V1, the layer IVB was labeled the most heavily, although in marmosets the upper tier of layer IVB stood out better and was not homogenously labeled as seen in the Old World monkeys; (ii) the infragranular layers of the MT, MT crescent, and parietal cortex contained the most numerous SMI-32-ir cells among the visual areas; (iii) the main chemoanatomical difference – apart from the different layer thickness – between V2 and V3 was the presence of SMI-32-ir neurons in layer II in the latter; (iv) the largest number layer VI pyramidal cells were found in the MT (middle temporal visual area) [6]. The described resemblances between the marmoset and the macaque in their SMI-32 chemoarchitecture further suggested that the New World monkey's and Old World monkey's visual cortices are very similar. The feline PMLS (posteromedial lateral suprasylvian cortex) also presented intense SMI 32 stain in layer V, giving further morphological proof of the functional kinship between the feline PMLS and the primate MT [14]. In general, the cat and especially the Old World and New World monkeys produced heavier labeling with the SMI 32 antibody in visual areas alongside the dorsal stream than they did in the ventral stream. In sum, the SMI 32 proved to be an excellent and incredibly powerful immunohistochemical marker for areal or nuclear parcellation not only in the cortex but in subcortical areas as well.

The other effective chemoanatomical marker of recent times is the cat-301. Contrary to the intracellularly staining SMI 32, it labels extracellular domains of the neuronal membranes. It can label pyramidal cells, but also non-pyramidal neurons of various forms [57]. A very similar cytoarchitectonic map of the visual cortical orchestra preceded the one made with the help of SMI 32 [27]. However, the latter marker gained wider use in chemoarchitectonic parcellation of the brain. In our material, this morphologically and chemically heterogeneous cortical cell population showed some overlap with the Nf-containing neurons, although no study has yet addressed this question specifically. It would certainly be of interest to learn whether cells exist with internal skeleton (Nfs) and external scaffolding (perineuronal nets) at the same time.

The developmental profile of SMI 32-ir neurons in human primary visual cortex (V1) has also been described [2]. This may serve as a useful morphological correlate of neuronal maturity, age determination and help for studying developmental diseases affecting the human isocortex. By now, a long series of neuropathological studies has utilized SMI 32-immunoreaction to detect cortical pyramidal cell loss or irregularities [103]. In addition, the maturation of the striate cortex of an Old World monkey [78] and a New World monkey [13] and its middle temporal area have all been described [14]. Essentially, the primate and human developmental profiles were very similar, only the timing was different.

In the primate primary visual cortex (V1), SMI 32-ir neurons distribute unequally in its functional segments (i.e. blobs and interblobs of layers II/III) [31], even though these pyramidal cells inside and outside of blobs do not differ in their soma size, spine density, or basal dendritic field structure [68]. Furthermore, SMI 32 staining of V1 produced a laminar pattern that was largely complementary to the staining pattern of cytochrome oxidase histochemistry (CO), a technique that specifically visualizes cortical blobs. These results suggest that there can be different sets of neurons within blobs and interblobs. Layer IV, however, receiving the strongest geniculate input, immunoreacted weakly for SMI 32 (Figure 14A). On the other hand, experimental interventions, such as monocular deprivation (MD) in the sensitive period of the visual system resulted in dramatic loss of Nfs in the deprived cortical ocular dominance column [32]. It seems that cytoskeletal changes contribute to MD-induced reorganization of neural connections within V1. Similarly, in experimentally strabismic monkeys, the number of SMI 32-ir pyramids was reduced in cortical columns belonging to the affected eye and absent at ocular dominance columnar borders [34].

Acknowledgements

The author was a scholar of the Hungarian Academy of Sciences (FK/96), the German Academic Exchange Service (DAAD) at the C. und O. Vogt Institut für Hirnforschung, Heinrich Heine Universität in Düsseldorf, Germany (#A/99/25615), and a postdoctoral fellow of the Fight for Sight Inc., NY at the Department of Anatomical Sciences and Neurobiology, University of Louisville, Kentucky (#PD03055). The author wishes to express his gratitude to Tünde Magyar, Lilia Igdalova and Lívia Herczeg for their expert technical assistance. Special thanks to Martin. J. Boyce and Lívia Herczeg for their useful comments and proof-reading. This study was also supported by the Hungarian Scientific Research Fund (T17282, CO/358A) and the National Pension Insurance Fund (#273-62484-9).

References

Al-Abdulla, N.A., Portera-Cailliau, C., Martin, L.J. (1998). Occipital cortex ablation in adult rat causes retrograde neuronal death in the lateral geniculate nucleus that resembles apoptosis. *Neuroscience*, 86(1):191-209.

Ang, L.C., Munoz, D.G., Shul, D., George, D.H. (1991). SMI-32 immunoreactivity in human striate cortex during postnatal development. *Brain Res Dev Brain Res.*, 61(1):103-109.

Ashwell, K.W., Paxinos, G. (2005). Cyto- and chemoarchitecture of the dorsal thalamus of the monotreme Tachyglossus aculeatus, the short beaked echidna. *J Chem Neuroanat.*, 30(4):161-183.

Ashwell, K.W., Zhang, L.L., Marotte, L.R. (2005). Cyto- and chemoarchitecture of the cortex of the tammar wallaby (Macropus eugenii): areal organization. *Brain Behav Evol.*, 66(2):114-136.

Baldauf, Z.B., Bickford, M.E. (2004). Comparison of cortical and subcortical glutamatergic innervation of the mammalian pretectum. *FENS Abstr.*, A122.1, vol. 2

Baldauf, Z.B. (2005). SMI-32 parcellates the visual cortical areas of the marmoset. *Neurosci Lett.*, 383(1-2):109-114.

Bickford, M.E., Guido, W., Godwin, D.W. (1998). Neurofilament proteins in Y-cells of the cat lateral geniculate nucleus: normal expression and alteration with visual deprivation. *J Neurosci.*, 18:6549–6557.

Black, M.M., Lasek, R.J. (1980). Slow components of axonal transport: two cytoskeletal networks. *J Cell Biol.*, 86(2):616-623.

Boire, D., Desgent, S., Matteau, I., Ptito, M. (2005). Regional analysis of neurofilament protein immunoreactivity in the hamster's cortex. *J Chem Neuroanat.*, 29(3):193-208.

Borostyánkői, Z.A., Görcs, T.J., Hámori, J. (1999). Immunocytochemical mapping of NPY and VIP neuronal elements in the cat subcortical visual nuclei, with special reference to the pretectum and accessory optic system. *Anat Embryol (Berl).*, 200(5):495-508.

Bourne, J.A., Rosa, M.G.P. (2003a). Neurofilament protein expression in the geniculostriate pathway of a New World monkey (Callithrix jacchus). *Exp Brain Res.*, 150:19–24.

Bourne, J.A., Rosa, M.G.P. (2003b). Laminar expression of neurofilament protein in the superior colliculus of the marmoset monkey (Callithrix jacchus). *Brain Res.,* 973(1):142-145.

Bourne, J.A., Warner, C.E., Rosa, M.G.P. (2005). Topographic and laminar maturation of striate cortex in early postnatal marmoset monkeys, as revealed by neurofilament immunohistochemistry. *Cereb Cortex* 15(6):740-748.

Bourne, J.A., Rosa, M.G.P. (2006). Hierarchical development of the primate visual cortex, as revealed by neurofilament immunoreactivity: early maturation of the middle temporal area (MT). *Cereb Cortex,* 16(3):405-414.

Brown, A. (1998). Contiguous phosphorylated and non-phosphorylated domains along axonal neurofilaments. *J Cell Sci.,* 111 (Pt 4):455-467.

Brown, M.E., Bridgman, P.C. (2004). Myosin function in nervous and sensory systems. *J Neurobiol.,* 58(1):118-130.

Campbell, M.J., Morrison, J.H. (1989). Monoclonal antibody to neurofilament protein (SMI-32) labels a subpopulation of pyramidal neurons in the human and monkey neocortex. *J Comp Neurol.,* 282:191–205.

Campbell, M.J., Hof, P.R., Morrison, J.H. (1991). A subpopulation of primate corticocortical neurons is distinguished by somatodendritic distribution of neurofilament protein. *Brain Res.,* 539:133–136.

Capani, F., Loidl, C.F., Pecci Saavedra, J. (1996). Unilateral enucleation induces an increase of 160 kd neurofilament in lateral geniculate nuclei synapses. *Biocell.,* 20(1):55-59.

Carriedo, S.G., Yin, H.Z., Lamberta, R., Weiss, J.H. (1995). In vitro kainate injury to large, SMI-32(+) spinal neurons is Ca2+ dependent. *Neuroreport,* 6(6):945-948.

Chaudhuri, A., Zangenehpour, S., Matsubara, J.A., Cynader, M.S. (1996). Differential expression of neurofilament protein in the visual system of the vervet monkey. *Brain Res.,* 709(1):17-26.

Clark, S.E., Garret, M., Platt, B. (2001). Postnatal alterations of GABA receptor profiles in the rat superior colliculus. *Neuroscience,* 104(2):441-454.

Coombs, J., van der List, D., Wang, G.Y., Chalupa, L.M. (2006). Morphological properties of mouse retinal ganglion cells. *Neuroscience,* 140(1):123-136.

Correa, C.L., da Silva, S.F., Lowe, J., Tortelote, G.G., Einicker-Lamas, M., Martinez, A.M., Allodi, S. (2004). Identification of a neurofilament-like protein in the protocerebral tract of the crab Ucides cordatus. *Cell Tissue Res.,* 318(3):609-615.

Cudkowicz, M., Kowall, N.W. (1990). Degeneration of pyramidal projection neurons in Huntington's disease cortex. *Ann Neurol.,* 27(2):200-204.

Datskovskaia, A., Eisenback, M.A., Bickford, M.E. (2002). *Synaptic organization of the cat perigeniculate nucleus. Program* No. 352.25. Washington, DC; Society for Neuroscience

DeYoe EA, Hockfield S, Garren H, Van Essen DC. (1990). Antibody labeling of functional subdivisions in visual cortex: Cat-301 immunoreactivity in striate and extrastriate cortex of the macaque monkey. *Vis Neurosci.,* 5(1):67-81.

Dieterich, D.C., Trivedi, N., Engelmann, R., Gundelfinger, E.D., Gordon-Weeks, P.R., Kreutz, M.R. (2002). Partial regeneration and long-term survival of rat retinal ganglion cells after optic nerve crush is accompanied by altered expression, phosphorylation and distribution of cytoskeletal proteins. *Eur J Neurosci.,* 15(9):1433-1443.

Drager, U.C., Hofbauer, A. (1984a). Antibodies to heavy neurofilament subunit detect a subpopulation of damaged ganglion cells in retina. *Nature*, 309(5969):624-626.

Drager, U.C., Edwards, D.L., Barnstable, C.J. (1984b). Antibodies against filamentous components in discrete cell types of the mouse retina. *J Neurosci.*, 4:2025–2042.

Duffy, K.R., Livingstone, M.S. (2003). Distribution of non-phosphorylated neurofilament in squirrel monkey V1 is complementary to the pattern of cytochrome-oxidase blobs. *Cereb Cortex.*, 13(7):722-727.

Duffy, K.R., Livingstone, M.S. (2005). Loss of neurofilament labeling in the primary visual cortex of monocularly deprived monkeys. *Cereb Cortex*, 15(8):1146-1154.

Eberhorn, A.C., Ardeleanu, P., Büttner-Ennever, J.A., Horn, A.K. (2005). Histochemical differences between motoneurons supplying multiply and singly innervated extraocular muscle fibers. *J Comp Neurol.*, 491(4):352-366.

Fenstemaker, S.B., Kiorpes, L., Movshon, J.A. (2001). Effects of experimental strabismus on the architecture of macaque monkey striate cortex. *J Comp Neurol.*, 438(3):300-317.

Fífková, E. (1985). Actin in the nervous system. *Brain Res.*, 356(2):187-215.

Fischer, I., Shea, T.B. (1991). Differential appearance of extensively phosphorylated forms of the high molecular weight neurofilament protein in regions of mouse brain during postnatal development. J Neuroimmunol., 31(1):73-81.

FitzGibbon, T., Solomon, S.G., Goodchild, A.K. (2000). Distribution of calbindin, parvalbumin, and calretinin immunoreactivity in the reticular thalamic nucleus of the marmoset: evidence for a medial leaflet of incertal neurons. *Exp Neurol.*, 164(2):371-383.

Friede, R.L., Samorajski, T. (1970). Axon caliber related to neurofilaments and microtubules in sciatic nerve fibers of rats and mice. *Anat Rec.*, 167(4):379-387.

Friede, R.L. (1972). Control of myelin formation by axon caliber (with a model of the control mechanism). *J Comp Neurol.*, 144(2):233-252.

Fuchs, E., Weber, K. (1994). Intermediate filaments: structure, dynamics, function, and disease. *Annu. Rev. Biochem.*, 63:345–382.

Fuchs, E., Cleveland, D.W. (1998). A structural scaffolding of intermediate filaments in health and disease. *Science*, 279:514–519.

Fuentes-Santamaria, V., Stein, B.E., McHaffie, J.G. (2006). Neurofilament proteins are preferentially expressed in descending output neurons of the cat the superior colliculus: A study using SMI-32. *Neuroscience*, 138(1):55-68.

Gábriel, R., Straznicky, C. (1992). Immunocytochemical localization of parvalbumin- and neurofilament triplet protein immunoreactivity in the cat retina: colocalization in a subpopulation of AII amacrine cells. *Brain Res.*, 595(1):133-136.

Gamlin, P.D. (2005). The pretectum: connections and oculomotor-related roles. *Prog Brain Res.*, 151:379-405.

Giolli, R.A., Blanks, R.H.I., Lui, F. (2005). The accessory optic system: basic organization with an update on connectivity, neurochemistry, and function. *Prog Brain Res.*, 151, pp. 407-440.

Goldstein, M.E., Cooper, H.S., Bruce, J., Carden, M.J., Lee, V.M., Schlaepfer, W.W. (1987). Phosphorylation of neurofilament proteins and chromatolysis following transection of rat sciatic nerve. *J Neurosci.*, 7(5):1586-1594.

Gotow, T., Tanaka, J. (1994). Phosphorylation of neurofilament H subunit as related to arrangement of neurofilaments. *J Neurosci Res.,* 37(6):691-713.

Graybiel, A.M., Berson, D.M. (1980). Histochemical identification and afferent connections of subdivisions in the lateralis posterior-pulvinar complex and related thalamic nuclei in the cat. *Neuroscience,* 5(7):1175-1238.

Griffin, J.W., George, E.B., Hsieh, S.T., Glass, J.D. (1995). Axonal degeneration and disorders of the axonal cytoskeleton, pages 375– 390. In: Waxman, S.G., Kocsis, J.D., Stys, P.K., eds. *The axon: structure, function and pathophysiology.* New York, Oxford University Press

van der Gucht, E., Burnat, K., Arckens, L. (2004). Characteristic expression of neurofilament protein defines six cortical areas and several subcortical divisions in mouse visual system, Program No. 300.7. San Diego, CA; Society for Neuroscience

van der Gucht, E., Vandesande, F., Arckens, L. (2001). Neurofilament protein: a selective marker for the architectonic parcellation of the visual cortex in adult cat brain, *J. Comp. Neurol.,* 441:345–368.

Gutierrez, C., Yaun, A., Cusick, C.G. (1995). Neurochemical subdivisions of the inferior pulvinar in macaque monkeys. *J Comp Neurol.,* 363(4):545-562.

Hajdu, F., Hässler, R., Somogyi, Gy., Wagner, A.. (1983). Neuronal and synaptic arrangements of the lateral geniculate nucleus in night-active primates. *Anat Embryol (Berl).* 168(3):341-348.

Hall, C.M. (1994). Differential regulation of two classes of neuronal intermediate filament proteins during optic nerve regeneration. *J Neurochem.,* 63(6):2307-2313.

Hartmann, H.A., Sun, D.Y. (1992). Regional mRNA changes in brain stem motor neurons from patients with amyotrophic lateral sclerosis. *Mol Chem Neuropathol.,* 17(3):249-257.

Hassiotis, M., Paxinos, G., Ashwell, K.W. (2004). Cyto- and chemoarchitecture of the cerebral cortex of the Australian echidna (Tachyglossus aculeatus). I. Areal organization. *J Comp Neurol.,* 475(4):493-517.

Hendry, S.H.C., Jones, E.G., Hockfield, S., McKay, R.D.G. (1988). Neuronal populations stained with the monoclonal antibody Cat-301 in the mammalian cerebral cortex and thalamus. *J Neurosci.,* 8(2):518-542.

Hockfield S. (1987). A Mab to a unique cerebellar neuron generated by immunosuppression and rapid immunization. *Science,* 237(4810):67-70.

Hockfield, S., Sur, M. (1990). Monoclonal antibody Cat-301 identifies Y-cells in the dorsal lateral geniculate nucleus of the cat. *J Comp Neurol.,* 300(3):320-330.

Hof, P.R., Glezer, I.I., Archin, N., Janssen, W.G., Morgane. P.J., Morrison, J.H. (1992). The primary auditory cortex in cetacean and human brain: a comparative analysis of neurofilament protein-containing pyramidal neurons. *Neurosci Lett.,* 146(1):91-95.

Hof, P.R., Morrison, J.H. (1995). Neurofilament protein defines regional patterns of cortical organization in the macaque monkey visual system: a quantitative immunohistochemical analysis. *J Comp Neurol.,* 352:161-186.

Hof, P.R., Bogaert, Y.E., Rosenthal, R.E., Fiskum, G. (1996). Distribution of neuronal populations containing neurofilament protein and calcium binding proteins in the canine neocortex: regional analysis and cell typology. *J. Chem. Neuroanat.,* 11:81–98.

Hoffman, P.N., Lasek, R.J. (1975). The slow component of axonal transport. Identification of major structural polypeptides of the axon and their generality among mammalian neurons. *J Cell Biol.*, 66(2):351-366.

Hoffman, P.N., Lasek, R.J. (1980). Axonal transport of the cytoskeleton in regenerating motor neurons: constancy and change. *Brain Res.*, 202(2):317-333.

Hoffman, P.N., Griffin, J.W., Gold, B.G., Price, D.L. (1985). Slowing of neurofilament transport and the radial growth of developing nerve fibers. *J Neurosci.*, 5(11):2920-2929.

Hong, F., Lee, W.H. (1991). Sequence similarity between part of human retinoblastoma susceptibility gene product and a neurofilament protein subunit. *Biosci Rep.*, 11(3):159-163.

Hornung, J.P., Riederer, B.M. (1999). Medium-sized neurofilament protein related to maturation of a subset of cortical neurons. *J Comp Neurol.*, 414(3):348-360.

Hubener, M., Bolz, J. (1992). Relationships between dendritic morphology and cytochrome oxidase compartments in monkey striate cortex. *J Comp Neurol.*, 324:67–80.

Hutsler, J.J., White, C.A., Chalupa, L.M. (1993). Neuropeptide Y immunoreactivity identifies a group of gamma-type retinal ganglion cells in the cat. *J Comp Neurol.*, 336(3):468-480.

Jaubert-Miazza, L., Green, E., Lo, F.S., Bui, K., Mills, J., Guido, W. (2005). Structural and functional composition of the developing retinogeniculate pathway in the mouse. *Vis Neurosci.*, 22(5):661-676.

Jin, L.Q., Zhang, G., Selzer, M.E. (2005). Lamprey neurofilaments contain a previously unreported 50-kDa protein. *J Comp Neurol.*, 483(4):403-414.

Kadota, T., Kadota, K. (1979). Filamentous contacts containing subjunctional dense lattice and tubular smooth endoplasmic reticulum in cat lateral geniculate nuclei. *Brain Res.* 177(1):49-59.

Kashiwagi, K., Ou, B., Nakamura, S., Tanaka, Y., Suzuki, M., Tsukahara, S. (2003). Increase in dephosphorylation of the heavy neurofilament subunit in the monkey chronic glaucoma model. *Invest Ophthalmol Vis Sci.*, 44(1):154-159.

King, C.E., Jacobs, I., Dickson, T.C., Vickers, J.C. (1997). Physical damage to rat cortical axons mimics early Alzheimer's neuronal pathology. *Neuroreport*, 8:1663-1665.

Kirkcaldie, M.T.K., Diskson, T.C., King, C.E., Grasby, D., Riederer, B.M., Vickers, J.C. (2002). Neurofilament triplet proteins are restricted to a subset of neurons in the rat neocortex. *J. Chem. Neuroanat.*, 24:163–171.

Kivela, T., Tarkkanen, A., Virtanen, I. (1986). Intermediate filaments in the human retina and retinoblastoma. An immunohistochemical study of vimentin, glial fibrillary acidic protein, and neurofilaments. *Invest Ophthalmol Vis Sci.*, 27(7):1075-1084.

Klooster, J., Vrensen, G.F., van der Want, J.J. (1995). Efferent synaptic organization of the olivary pretectal nucleus in the albino rat. *An ultrastructural tracing study. Brain Res.*, 688(1-2):47-55.

Kogan, C.S., Zangenehpour, S., Chaudhuri, A. (2000). Developmental profiles of SMI-32 immunoreactivity in monkey striate cortex. *Brain Res Dev Brain Res.*, 119(1):85-95.

Kong, J., Tung, V.W., Aghajanian, J., Xu, Z. (1998). Antagonistic roles of neurofilament subunits NF-H and NF-M against NF-L in shaping dendritic arborization in spinal motor neurons. *J Cell Biol.*, 140(5):1167-1176.

Lasek, R.J., Hoffman, P.N. (1976). The neuronal cytoskeleton, axonal transport and axonal growth. Cold Spring Harbor Conf. *Cell Prolif.,* 3: 1021-1049.

Lasek, R.J., Black, M.M. (1986). Intrinsic determinants of neuronal form and function. Neurology and Neurobiology, Vol 37. A. Liss, NY (Proceedings of a Meeting on Intrinsic Determinants of Neuronal Form and Function, Bioarchitectonics Center, School of Medicine, Case Western Reserve University, Cleveland, OH)

Laskawi, R., Wolff, J.R. (1996). Changes in the phosphorylation of neurofilament proteins in facial motoneurons following various types of nerve lesion. *ORL J Otorhinolaryngol Relat Spec.,* 58(1):13-22.

Lee, V.M., Carden, M.J., Schlaepfer, W.W. (1986). Structural similarities and differences between neurofilament proteins from five different species as revealed using monoclonal antibodies. *J Neurosci.,* 6(8):2179-2186.

Lee, M.K., Cleveland, D.W. (1996). Neuronal intermediate filaments. *Annu Rev Neurosci.,* 19:187-217.

Lieberman, A.R., Spacek, J. (1997). Filamentous contacts: the ultrastructure and three-dimensional organization of specialized non-synaptic interneuronal appositions in thalamic relay nuclei. *Cell Tissue Res.,* 288(1):43-57.

Liu, Y., Dyck, R., Cynader, M. (1994). The correlation between cortical neuron maturation and neurofilament phosphorylation: a developmental study of phosphorylated 200 kDa neurofilament protein in cat visual cortex. *Brain Res Dev Brain Res.,* 81(2):151-161.

Liu, Q., Xie, F., Siedlak, S.L., Nunomura, A., Honda, K., Moreira, P.I., Zhua, X., Smith, M.A., Perry, G. (2004). Neurofilament proteins in neurodegenerative diseases. *Cell Mol Life Sci.,* 61(24):3057-3075.

MacDonald, V., Halliday, G. (2002). Pyramidal cell loss in motor cortices in Huntington's disease. *Neurobiol Dis.,* 10(3):378-386.

Maler, L., Leclerc, N., Hawkes, R. (1986). A monoclonal antibody to mammalian neurofilament protein stains somata and dendrites in gymnotid fish. Brain Res., 378(2):337-346.

Marani, E., Choufoer, H., van der Veeken, J. (1990). The neurofilament architecture of the rat suprachiasmatic nucleus. *Eur J Morphol.,* 28(2-4):279-288.

Matus, A, (1988). Neurofilament protein phosphorylation -where, when and why. *Trends Neurosci.,* 11(7):291-292.

May, P.J. (2005). The mammalian superior colliculus: laminar structure and connections. *Prog Brain Res.,* 151:321-378.

McKay, R., Johansen, J., Hockfield, S. (1984). Monoclonal antibody identifies a 63,000 dalton antigen found in all central neuronal cell bodies but in only a subset of axons in the leech. *J Comp Neurol.,* 226(3):448-455.

Meller, D., Schmidt-Kastner, R., Eysel, U.T. (1993). Immunohistochemical studies on neurofilamentous hypertrophy in degenerating retinal terminals of the olivary pretectal nucleus in the rat. *J Comp Neurol.,* 331(4):531-539.

Meller, D., Eysel, U.T., Schmidt-Kastner, R. (1994). Transient immunohistochemical labelling of rat retinal axons during Wallerian degeneration by a monoclonal antibody to neurofilaments. *Brain Res.,* 648(1):162-166.

Molnár, Z., Cheung, A.F. (2006). Towards the classification of subpopulations of layer V pyramidal projection neurons. *Neurosci Res.*, 55(2):105-15.

Motil, J., Chan, W.K., Dubey, M., Chaudhury, P., Pimenta, A., Chylinski, T.M., Ortiz, D.T., Shea, T.B. (2006). Dynein mediates retrograde neurofilament transport within axons and anterograde delivery of NFs from perikarya into axons: Regulation by multiple phosphorylation events. *Cell Motil Cytoskeleton*, 63(5):266-286.

Narisawa, Y., Hashimoto, K., Kohda, H. (1994). Immunohistochemical demonstration of the expression of neurofilament proteins in Merkel cells. *Acta Derm Venereol.*, 74(6):441-443.

Neuhoff, H., Sassoe-Pognetto, M., Panzanelli, P., Maas, C., Witke, W., Kneussel, M. (2005). The actin-binding protein profilin I is localized at synaptic sites in an activity-regulated manner. *Eur J Neurosci.*, 21(1):15-25.

Novotny, G.E. (1979). Observations on the lateral geniculate nucleus of the monkey (Macaca fascicularis) after eye removal: a light and electron microscopic study. I. Classification and degeneration of optic fibre terminals. *J Hirnforsch.*, 20(6):561-580.

Paxinos, G., Carrive, P., Wang, H., Wang, P.-Y. (1999a). *Chemoarchitectonic atlas of the rat brainstem*, Academic Press, San Diego

Paxinos, G., Kus, L., Ashwell, K., Watson, C. (1999b). *Chemoarchitectonic atlas of the rat forebrain*, Academic Press, San Diego

Petzold, A. (2005). Neurofilament phosphoforms: surrogate markers for axonal injury, degeneration and loss. *J Neurol Sci.*, 233(1-2):183-198.

Pisu, M.B., Scherini, E., Bernocchi, G. (1998). Immunocytochemical changes of cytoskeleton components and calmodulin in the frog cerebellum and optic tectum during hibernation. *J Chem Neuroanat.*, 15(2):63-73.

Pleasure, S.J., Selzer, M.E., Lee, V.M. (1989). Lamprey neurofilaments combine in one subunit the features of each mammalian NF triplet protein but are highly phosphorylated only in large axons. *J Neurosci.*, 9(2):698-709.

Prasad, S.S., Schnerch, A., Lam, D.Y., To, E., Jim, J., Kaufman, P.L., Matsubara, J.A. (2002). Immunohistochemical investigations of neurofilament M' and alphabeta-crystallin in the magnocellular layers of the primate lateral geniculate nucleus. *Brain Res Mol Brain Res.*, 109(1-2):216-220.

Preuss, T.M., Qi, H., Kaas, J.H. (1999). Distinctive compartmental organization of human primary visual cortex. *Proc Natl Acad Sci U S A.*, 96(20):11601-11606.

Prince, H.E. (2005). Biomarkers for diagnosing and monitoring autoimmune diseases. *Biomarkers.*, 10 Suppl 1:S44-49.

Riederer, B.M., Porchet, R., Marugg, R., Binder, L.I. (1993). Solubility of cytoskeletal proteins in immunohistochemistry and the influence of fixation. *J. Histochem. Cytochem.*, 41:609-616.

Sasaki, S., Maruyama, S. (1994). Immunocytochemical and ultrastructural studies of the motor cortex in amyotrophic lateral sclerosis. *Acta Neuropathol (Berl).*, 87:578–585.

Shah, J.V., Flanagan, L.A., Janmey, P.A., Leterrier, J.F. (2000). Bidirectional translocation of neurofilaments along microtubules mediated in part by dynein/dynactin. *Mol Biol Cell.*, 11(10):3495-3508.

Shaw, G., Debus, E., Weber, K. (1984). The immunological relatedness of neurofilament proteins of higher vertebrates. *Eur J Cell Biol.*, 34(1):130-136.

Shaw, G. (1991). Neurofilament proteins. In *The Neuronal Cytoskeleton.* R.D. Burgoyne, ed. Wiley-Liss, Inc., New York. pp 621–635.

Shea, T.B., Flanagan, L.A. (2001). Kinesin, dynein and neurofilament transport. *Trends Neurosci.*, 24(11):644-648.

Shea, T.B., Jung, C., Pant, H.C. (2003). Does neurofilament phosphorylation regulate axonal transport? *Trends Neurosci.*, 26(8):397-400.

Soares, J.G., Gattass, R., Souza, A.P., Rosa, M.G., Fiorani, M. Jr., Brandao, B.L. (2001). Connectional and neurochemical subdivisions of the pulvinar in Cebus monkeys.*Vis Neurosci.*, 18(1):25-41.

Steriade, M. (2005). Sleep, epilepsy and thalamic reticular inhibitory neurons. *Trends Neurosci.*, 28(6):317-324.

Sternberger, L.A., Sternberger, N.H. (1983). Monoclonal antibodies distinguish phosphorylated and nonphosphorylated forms of neurofilaments in situ. *Proc Natl Acad Sci USA*, 80:6129-6130.

Straznicky, C., Vickers, J.C., Gábriel, R., Costa, M. (1992). A neurofilament protein antibody selectively labels a large ganglion cell type in the human retina. *Brain Res.*, 582(1):123-128.

Szaro, B.G., Gainer, H. (1988). Identities, antigenic determinants, and topographic distributions of neurofilament proteins in the nervous systems of adult frogs and tadpoles of Xenopus laevis. *J Comp Neurol.*, 273(3):344-358.

Taniguchi, T., Shimazawa, M., Hara, H. (2004). Alterations in neurofilament light in optic nerve in rat kainate and monkey ocular hypertension models. *Brain Res.*, 1013(2):241-248.

Tesser, P., Jones, P.S., Schechter, N. (1986). Elevated levels of retinal neurofilament mRNA accompany optic nerve regeneration. *J Neurochem.*, 47(4):1235-1243.

Tsang, Y.M., Chiong, F., Kuznetsov, D., Kasarskis, E., Geula, C. (2000). Motor neurons are rich in non-phosphorylated neurofilaments: cross-species comparison and alterations in ALS. *Brain Res.*, 861(1):45-58.

Ulfig, N., Nickel, J., Bohl, J. (1998). Transient features of the thalamic reticular nucleus in the human foetal brain. *Eur J Neurosci.*, 10(12):3773-3784.

Xu, Z., Dong, D.L.-Y., Cleveland, D.W. (1994). Neuronal intermediate filaments: new progress on an old subject. *Curr Opin Neurobiol.*, 4:655–661.

Xu, Z., Marszalek, J.R., Lee, M.K., Wong, P.C., Folmer, J., Crawford, T.O., Hsieh, S.I., Griffin, J.W., and Cleveland, D.W. (1996). Subunit composition of neurofilaments specifies axonal diameter. *J. Cell Biol.*, 133, 1061-1069.

Yasuda, Y., Fujita, S. (2003). Distribution of MAP1A, MAP1B, and MAP2A&B during layer formation in the optic tectum of developing chick embryos. *Cell Tissue Res.*, 314(3):315-324.

Vickers, J.C., Hof, P.R., Schumer, R.A., Wang, R.F., Podos, S.M., Morrison, J.H. (1997). Magnocellular and parvocellular visual pathways are both affected in a macaque monkey model of glaucoma. *Aust N Z J Ophthalmol.*, 25(3):239-243.

Voelker, C.C., Garin, N., Taylor, J.S., Gähwiler, B.H., Hornung, J.P., Molnár, Z. (2004). Selective neurofilament (SMI-32, FNP-7 and N200) expression in subpopulations of layer V pyramidal neurons in vivo and in vitro. *Cereb. Cortex* 14:1276–1286.

Völgyi, B., Bloomfield, S.A. (2002). Axonal neurofilament-H immunolabeling in the rabbit retina. *J Comp Neurol.*, 453(3):269-279.

Walsh, I., Marani, E., van der Berg, R.J., Rietveld, W.J. (1990). The suprachiasmatic nucleus of the rat hypothalamus in culture: an anatomical and electrophysiological study. *Eur J Morphol.*, 28(2-4):317-329.

van der Want, J.J., Nunes Cardozo, J.J., van der Togt, C. (1992). GABAergic neurons and circuits in the pretectal nuclei and the accessory optic system of mammals. *Prog Brain Res.*, 90:283-305.

Zhang, J.H., Sampogna, S., Morales, F.R., Chase, M.H. (2000). Age-dependent changes in the midsized neurofilament subunit in sensory-motor systems of the cat brainstem: an immunocytochemical study. *J Gerontol A Biol Sci Med Sci.*, 55(5):B233-241.

Zhang, Z., Casey, D.M., Julien, J.P., Xu, Z. (2002). Normal dendritic arborization in spinal motoneurons requires neurofilament subunit L. *J Comp Neurol.*, 450(2):144-152.

Zheng, Y.L., Li, B.S., Veeranna, Pant, H.C. (2003). Phosphorylation of the head domain of neurofilament protein (NF-M): a factor regulating topographic phosphorylation of NF-M tail domain KSP sites in neurons. *J Biol Chem.*, 278(26):24026-24032.

Zhao, Y., Szaro, B.G. (1995). The optic tract and tectal ablation influence the composition of neurofilaments in regenerating optic axons of Xenopus laevis. *J Neurosci.*, 15(6):4629-4640.

Index

A

acceptance, 105, 134
accumulation, viii, ix, xi, 10, 25, 27, 30, 31, 32, 36, 37, 38, 40, 44, 47, 49, 54, 63, 69, 75, 76, 77, 81, 83, 85, 86, 87, 96, 125, 130, 131, 135, 146
acid, 28, 34, 38, 42, 56, 118, 122, 130
action potential, 3, 67, 83, 105
activation, 8, 17, 23, 37, 46, 50, 88, 92, 116, 125, 126, 128
adduction, 50
adhesion, 14
adulthood, viii, 25, 27, 66
affect, x, 37, 48, 64, 76, 116, 134
afferent nerve, 2
age, ix, 32, 35, 36, 37, 54, 64, 65, 68, 94, 122, 124, 138, 153
ageing, 2
agent, 32
aggregates, 30, 37, 59, 62, 67, 71, 104, 128
aggregation, 32, 37, 39, 40, 67
aging, 63, 74
alimentary canal, 100
ALS, 10, 27, 32, 33, 34, 35, 39, 41, 47, 50, 51, 63, 66, 70, 78, 134, 161
alternative, vii, 1, 130
alternatives, 23
alters, 23, 31
Alzheimer's disease, 19, 44, 47, 49, 50, 77
amino acids, 28, 59
ammonia, 45
ammonium, 38, 41

amyotrophic lateral sclerosis, viii, 10, 19, 25, 27, 33, 34, 35, 40, 41, 43, 44, 47, 49, 50, 51, 63, 66, 70, 71, 72, 74, 76, 77, 134, 157, 160
anatomy, 112
animal welfare, 101
animals, xi, 22, 89, 101, 102, 116, 122
ANOVA, 15, 16
anoxia, 40
antibody, x, 8, 9, 11, 12, 77, 84, 93, 115, 118, 119, 120, 121, 122, 123, 124, 126, 130, 134, 137, 150, 153, 155, 157, 159, 161
antigen, 159
apoptosis, 31, 84, 91, 94, 128, 154
aspartate, 86, 93
association, 20, 28, 32, 33, 34, 35, 36, 45, 46, 51, 74, 140
astrocytes, 45, 49, 55, 119
astrogliosis, ix, 81, 85
ataxia, 65, 94
atrophy, ix, 36, 37, 38, 39, 43, 44, 54, 65, 68, 71, 83, 90, 116, 117, 124, 129
attacks, 83
attention, 133
auditory cortex, 3, 157
auditory nerve, 20, 21, 22
Australia, 1
autoantibodies, 96
autoimmune disease, 160
autoimmunity, 84
autonomic nervous system, 100, 113
autosomal dominant, 32, 33, 94
autosomal recessive, 36
availability, 35
axon terminals, 133

Index

axonal degeneration, 34, 39, 40, 50, 83, 86, 87, 88, 90
axonal pathology, ix, 81, 83, 84, 87
axons, vii, viii, ix, x, xi, 1, 2, 5, 10, 13, 15, 16, 22, 23, 25, 29, 30, 31, 36, 37, 43, 45, 46, 48, 49, 51, 53, 54, 58, 60, 61, 62, 63, 64, 67, 68, 70, 71, 74, 75, 76, 77, 78, 79, 81, 83, 85, 86, 87, 88, 89, 90, 91, 92, 93, 94, 95, 96, 97, 100, 101, 103, 104, 105, 108, 109, 110, 115, 117, 123, 124, 126, 127, 128, 130, 131, 133, 134, 135, 136, 141, 158, 159, 160, 162

B

basal forebrain, 91, 94
basal ganglia, 38
basilar membrane, 3, 8, 10, 11, 12, 13
behavior, 75
bile, 100
bile duct, 100
binding, 8, 33, 48, 61, 65, 70, 71, 105, 112, 133, 157, 160
birth, 5
bladder, 100
blocks, 96
blood, ix, 38, 50, 99, 116, 118, 126
blood vessels, ix, 99
body, 50, 65, 66, 68, 73, 77, 102, 110, 117
brain, vii, x, xi, 2, 3, 11, 22, 23, 28, 32, 33, 34, 35, 36, 38, 39, 40, 41, 44, 46, 49, 62, 77, 82, 83, 86, 91, 115, 116, 117, 119, 122, 124, 125, 126, 127, 128, 129, 130, 133, 134, 137, 150, 151, 153, 156, 157, 161
brain damage, 38, 39, 40
brain growth, 129
brain stem, 157
brainstem, xi, 3, 17, 20, 39, 131, 137, 138, 140, 160, 162
branching, 133, 148, 150
Brazil, 129
breakdown, 8, 10, 18
buffer, 118

C

calcitonin, x, 100, 101, 104, 109, 111, 113
calcium, 8, 21, 50, 85, 95, 105, 112, 133, 157

caliber, viii, x, xi, 19, 20, 30, 31, 43, 47, 48, 49, 51, 53, 58, 62, 63, 68, 71, 72, 73, 75, 76, 96, 97, 115, 116, 117, 124, 125, 126, 128, 131, 133, 156
candidates, 32
carbonyl groups, 31
cardiac arrest, viii, 26, 27, 39, 40, 49
cardiac muscle, 55
caspases, 19
cast, 46
categorization, 103
cell, vii, viii, ix, x, xi, 1, 2, 3, 4, 5, 6, 9, 10, 11, 14, 17, 18, 20, 21, 25, 26, 29, 30, 38, 41, 45, 46, 47, 54, 57, 58, 59, 60, 61, 62, 63, 64, 65, 66, 68, 74, 75, 82, 84, 87, 89, 91, 92, 93, 94, 96, 97, 99, 100, 101, 115, 116, 117, 119, 120, 122, 123, 124, 125, 126, 129, 133, 134, 135, 143, 145, 146, 150, 151, 153, 156, 157, 159, 161
cell body, 5, 6, 20, 26, 29, 59, 61, 62, 64, 65, 66, 68, 124, 126, 133
cell culture, x, xi, 41, 115, 116, 117, 119, 120, 122, 123, 124, 125, 126, 129
cell death, ix, 54, 87, 93
cell line, 59, 91, 94, 96, 97
cell surface, 14, 46, 58, 74
central nervous system, viii, ix, 25, 27, 36, 81, 82, 83, 84, 86, 89, 90, 91, 94, 97, 117, 124, 127, 134, 135
cerebellum, 39, 160
cerebral cortex, 22, 42, 50, 58, 72, 157
cerebrospinal fluid, viii, 26, 27, 38, 39, 41, 82, 86, 127, 150
channel blocker, 86
channels, 83, 85, 94
chicken, 77, 78, 140
CHO cells, 127
cholinesterase, 148
choroid, 126
chromosome, 28, 58
chronic glaucoma, 46, 158
classes, 55, 157
classification, 102, 103, 105, 111, 113, 160
clusters, 85, 135, 136
CNS, 26, 28, 29, 30, 35, 36, 38, 39, 46, 55, 69, 75, 82, 85, 86, 89, 95
cochlea, vii, 1, 2, 3, 4, 5, 6, 7, 9, 10, 11, 12, 13, 14, 15, 16, 17, 18, 19, 20, 21, 22, 23
cochlear implant, vii, 1, 17, 18, 22
coding, x, 34, 99, 103, 104, 106, 107
codon, 67
cognition, xi, 116, 117

cognitive function, 32, 127
cohort, 133
colon, 106, 111, 112, 113, 114
communication, 150
competition, 30, 145
complement, 84
complexity, 82
complications, 40
components, viii, ix, 2, 3, 17, 25, 31, 37, 39, 55, 89, 90, 99, 101, 102, 154, 156, 160
composition, viii, 53, 133, 158, 161, 162
concentration, x, 39, 115, 117, 118, 120, 125
conduct, 105
conduction, viii, ix, 17, 25, 30, 33, 37, 47, 54, 58, 65, 69, 76, 78, 83, 85, 96, 116, 117, 125, 126, 134
configuration, 13
connective tissue, 102
connectivity, 156
construction, 118
control, 2, 12, 15, 16, 49, 67, 76, 93, 96, 124, 156
control group, 15
copper, 44
correlation, 20, 58, 87, 88, 159
cortex, 132, 135, 150, 151, 153, 154, 155, 156, 157, 158, 159, 160
cortical neurons, 23, 42, 64, 97, 128, 158
coupling, 31, 101
creatine, 38, 41
critical period, 39
Crohn's disease, 111
CSF, 39, 40, 49, 50, 82, 86
cues, 14
culture, 47, 62, 88, 92, 96, 119, 123, 162
cytoarchitecture, 15, 45
cytochrome, 132, 153, 156, 158
cytokines, 83, 84, 114
cytokinesis, 26
cytoplasm, 2, 37
cytosine, 119

D

database, 84
death, vii, 1, 2, 18, 32, 36, 66, 87, 91, 154
defects, ix, 37, 38, 54, 75, 81, 89
deficiency, xi, 38, 39, 49, 75, 78, 116, 124, 126, 127, 129, 135
deficit, 39, 94
definition, 54, 68, 107
degenerate, 86, 96

degradation, vii, 1, 8, 17, 22, 31, 37, 48, 67, 70, 133
delivery, 16, 92, 160
dementia, ix, 36, 37, 49, 54, 63, 66, 70, 116, 117, 125
demyelinating disease, 84
demyelination, vii, 1, 2, 9, 10, 11, 13, 15, 36, 84, 87, 88, 90, 93, 94, 96
dendrites, vii, viii, 29, 30, 53, 55, 102, 103, 104, 105, 109, 133, 134, 136, 137, 138, 140, 142, 144, 145, 147, 148, 149, 150, 159
dendritic arborization, 21, 29, 46, 51, 134, 158, 162
density, 9, 15, 17, 48, 85, 86, 89, 95, 102, 119, 153
dephosphorylation, ix, 8, 17, 29, 31, 36, 46, 64, 81, 85, 87, 93, 158
depolarization, 3, 22
depolymerization, viii, 25
deposition, 84
deposits, 67
deprivation, 19, 92, 132, 153, 154
deregulation, 32
derivatives, 55
desensitization, 46, 58, 74
destruction, 86, 87, 88, 92
detection, xi, 86, 120, 127, 131, 133, 150
diabetes, vii, x, 23, 115, 116, 117, 118, 125, 127, 129
diabetic neuropathy, viii, 25, 27, 37, 43, 47, 63, 66, 71, 116, 128, 134
diabetic patients, 37
differentiation, 5, 14, 23, 28, 43, 93, 96, 104, 117, 125, 126, 129, 146, 148
diphtheria, 128
direct observation, 59
disability, ix, 81, 82, 83, 84, 86, 89, 90, 93, 94
discrimination, x, 100
disease activity, 39
disease progression, 84, 85, 92, 95
disease rate, 84
disorder, 33, 35, 36, 82
dissociation, viii, 53, 84
distribution, x, xi, 20, 36, 48, 73, 85, 94, 115, 117, 126, 127, 131, 135, 136, 137, 138, 140, 148, 150, 151, 152, 155
diuretic, 9
diversity, 40
division, 100, 117, 132, 147, 149
DNA, 56, 84
domain, ix, xi, 17, 20, 28, 30, 32, 35, 37, 42, 43, 45, 46, 47, 51, 54, 56, 57, 58, 61, 62, 64, 66, 68, 77, 78, 116, 125, 126, 133, 162

dominance, 153
donors, 110
dopamine, 31, 35, 46, 58, 74
draft, 72
drugs, 83, 95
duodenum, 112
duplication, 55, 58
duration, 86

E

edema, 38, 39, 44
EEG, 116
electron microscopy, 54
electrophoresis, 118
embryo, 23, 41, 42, 127
embryogenesis, 6, 41
encephalomyelitis, 84, 95
encephalopathy, 39, 44
encoding, 36, 41, 43, 70
encouragement, 69
endothelial cells, 128
energy, 61, 64
energy efficiency, 61
England, 118, 120
environment, 13, 28, 83, 88
enzymes, 103
epilepsy, 161
epithelial cells, 55
epithelium, 134
equilibrium, 126
etiology, 116, 124
evidence, ix, 13, 14, 15, 38, 39, 43, 46, 54, 71, 74, 75, 81, 82, 83, 84, 85, 86, 87, 88, 89, 90, 94, 99, 107, 112, 126, 156
evolution, 61, 64, 74
excitability, 133
excitation, 150
excitotoxicity, 11, 37, 93
exclusion, 39
exposure, 2, 38
expression, vii, viii, 2, 5, 6, 8, 9, 10, 12, 14, 15, 18, 19, 20, 25, 27, 28, 31, 36, 38, 39, 40, 41, 42, 46, 49, 51, 58, 71, 74, 77, 78, 91, 94, 95, 96, 103, 114, 127, 128, 146, 154, 155, 157, 160, 162
extrapolation, 106

F

failure, 14, 18, 83, 85, 92
family, 2, 3, 41, 44, 70, 72, 91, 96, 117
family members, 96
feedback, 8, 91
feet, 33
females, 90
fibers, 23, 26, 28, 29, 76, 133, 135, 136, 138, 156
fibroblasts, 42, 55
filament, vii, 2, 3, 15, 17, 26, 28, 29, 37, 41, 42, 43, 45, 46, 48, 51, 54, 55, 58, 61, 62, 64, 67, 68, 70, 71, 72, 73, 75, 78, 82, 125, 134, 157
fish, 135, 140, 159
fixation, 160
flexibility, 6
fluid, 5, 39, 59, 86, 125, 130
fluorescence, 29, 78
focusing, 83
foramen, 8
forebrain, 48, 160
formaldehyde, 120
France, 118
functional analysis, 48
functional changes, 17

G

ganglion, 3, 4, 16, 19, 20, 21, 22, 23, 47, 50, 59, 60, 63, 70, 71, 75, 78, 92, 96, 102, 104, 109, 135, 155, 156, 158, 161
gastrointestinal tract, ix, 99, 112, 113
gene, vii, viii, x, 14, 20, 21, 23, 28, 31, 32, 34, 35, 36, 41, 43, 44, 45, 46, 47, 48, 50, 51, 53, 54, 56, 58, 59, 63, 65, 66, 70, 71, 72, 73, 74, 75, 76, 78, 79, 89, 100, 101, 104, 109, 111, 113, 114, 127, 128, 129, 133, 158
gene expression, 23, 58, 73, 127, 128, 129, 133
gene therapy, 20
gene transfer, 14, 21
generation, 3, 150
genes, viii, 10, 25, 27, 28, 32, 33, 41, 43, 44, 53, 55, 56, 57, 58, 72, 127
genetic mutations, 40, 66
genetics, 2, 61
Germany, 99, 154
gestation, 5, 6
gestational age, x, 115, 119, 123
glaucoma, 36, 135, 145, 161

glia, 41, 93, 114, 122, 141
glial cells, viii, 25, 26, 55, 100, 101, 111
glucose, 118, 122, 126, 128
glutamate, 37, 45, 91, 93
glutamic acid, 56
glycogen, xi, 44, 116, 128, 129
glycosylation, 27, 28, 29, 31, 33, 46, 56
goals, 18, 88
grants, 40, 110
grouping, 83, 135
groups, 15, 100
growth, vii, ix, x, 2, 5, 6, 10, 12, 13, 15, 17, 18, 23, 28, 29, 30, 31, 38, 41, 48, 49, 54, 58, 62, 69, 72, 73, 76, 82, 88, 89, 90, 91, 92, 93, 94, 96, 115, 117, 119, 124, 126, 127, 128, 129, 134, 138, 158, 159
growth dynamics, 23
growth factor, 23, 82, 88, 89, 90, 91, 92, 93, 94, 96, 117, 119, 128
guidance, 14, 26, 40, 55
Guinea, 12
gut, ix, x, 99, 100, 101, 102, 103, 107, 111, 113

H

hair cells, vii, 1, 2, 3, 4, 6, 7, 8, 9, 10, 14, 18, 21, 22
hands, 33
health, 43, 72, 73, 156
heat, 37, 40, 67, 69, 71
hemisphere, 150
hepatic encephalopathy, viii, 25, 27
heterogeneity, 84, 96
histochemistry, 132, 153
histology, 18, 133
homeostasis, 22, 32, 76
Honda, 21, 46, 74, 159
human brain, 27, 87, 157
human development, 6, 153
human genome, 72
Hungary, 131
Huntington's disease, 37, 43, 155, 159
hyperplasia, 146
hypertension, 135, 161
hypertrophy, 159
hypoglossal nerve, 23
hypothalamus, 127, 132, 136, 162
hypothesis, 5, 29, 64, 66, 86, 103

I

identification, x, 45, 100, 103, 105, 107, 140, 143, 145, 157
identity, 26
IFM, x, 115
ileum, 104, 106, 110, 111, 112
images, 140, 148, 151
immune system, 101
immunization, 157
immunogenicity, 71
immunoglobulin, 84
immunohistochemistry, x, 12, 49, 100, 103, 105, 107, 110, 114, 115, 119, 138, 155, 160
immunoreactivity, 5, 6, 9, 20, 70, 104, 105, 106, 111, 112, 113, 114, 137, 149, 154, 155, 156, 158
immunosuppression, 157
impregnation, 105
in situ hybridization, 41, 129
in vitro, 9, 14, 41, 42, 87, 91, 95, 97, 125, 129, 162
inclusion, ix, 37, 42, 50, 54, 63, 66, 70, 73, 75
induction, 111
infants, 38
infection, 2
inflammation, ix, 81, 83, 85, 87, 88, 93
inflammatory bowel disease, 111
inflammatory cells, 85
inflammatory demyelination, 94
inflammatory disease, 85
influence, ix, 11, 14, 17, 29, 62, 81, 86, 89, 90, 91, 92, 134, 145, 160, 162
inhibition, 31, 45, 87, 88, 92, 93, 117, 125, 126
inhibitor, x, 47, 115, 117, 118, 120, 124, 126
inner ear, 5, 19, 22, 23
input, 105, 145, 153
insertion, 35, 50, 67
insight, 86
insulin, x, xi, 37, 49, 82, 91, 96, 115, 116, 117, 118, 119, 120, 122, 123, 124, 125, 126, 127, 128, 129, 130
insulin dependent diabetes, 116, 124
insulin resistance, 129
insulin signaling, 130
integration, 61, 68
integrity, vii, 2, 17, 31, 37, 86, 134
intensity, 140, 142
interaction, 78, 92, 97
interactions, viii, 2, 5, 14, 25, 26, 28, 30, 31, 36, 61, 90, 95, 96, 97
interest, 110, 153

interface, 18
interferon, 83, 95
interneurons, ix, 99, 103, 106, 113, 145, 146, 148
internode, 30
interpretation, 88
intervention, 17, 22
intestine, 108, 109, 112, 113
intracerebral hemorrhage, 48
invaginate, 5
ischemia, 39
isolation, 54

K

keratin, 55, 71, 72, 134
knowledge, ix, 69, 93, 99, 135, 136

L

labeling, 21, 29, 59, 136, 138, 143, 148, 153, 155, 156
laminar, 153, 155, 159
lamination, 146
lateral sclerosis, 41, 67, 70, 72
lead, ix, 2, 17, 18, 34, 37, 38, 64, 67, 68, 82, 86, 87, 88, 90
lens, 55
lesions, ix, 38, 54, 67, 81, 83, 84, 85, 86, 87, 88, 90, 92, 94, 95, 96
life span, 89
lifetime, 31
links, ix, 99, 105
lipid peroxidation, 50
lipids, 95
local government, 101
localization, x, 19, 20, 21, 40, 78, 116, 120, 123, 124, 126, 128, 130, 156
location, 75, 100
locus, 128
long distance, 137
lumen, x, 100
lying, 102, 143, 148
lymphocytes, 84
lysine, 31, 87
lysis, 118

M

machinery, 58

macrophages, 84, 85
magnetic resonance, 83, 86, 95
magnetic resonance spectroscopy, 86, 95
mammalian brain, 44, 134
mantle, 149
mapping, 154
mass, 42, 129, 144, 148
mass spectrometry, 42, 129
maturation, vii, 2, 5, 23, 28, 31, 39, 51, 63, 74, 79, 93, 95, 128, 130, 136, 153, 155, 158, 159
MBP, 82, 89
measurement, 15
mechanical stress, 26
membranes, 153
memory, 36
men, 36
mental retardation, 38, 39
mentor, 69
messenger RNA, 47
metabolism, vii, 22, 38, 39
metabolites, 38, 86
mice, viii, xi, 10, 13, 15, 23, 43, 44, 45, 48, 51, 53, 54, 56, 57, 58, 59, 62, 63, 64, 65, 70, 71, 72, 73, 76, 79, 89, 90, 93, 95, 116, 117, 118, 121, 122, 124, 126, 127, 128, 130, 156
microheterogeneity, 72
microinjection, 77
microphotographs, 123
microscope, 120
microscopy, 7, 18, 63
midbrain, 137, 140
migration, 56
minority, 106, 107, 108, 109
mitochondria, 61
mitogen, x, 17, 23, 45, 95, 97, 115, 129
mitosis, 11, 26
MMA, 34, 35, 38
mode, 61, 83, 125
models, 9, 10, 32, 38, 61, 72, 76, 86, 87, 96, 135, 161
MOG, 82, 89
mole, 30
molecular biology, 54
molecular mass, 42, 46, 48, 72, 74, 76, 77
molecular medicine, 76
molecular weight, viii, xi, 3, 4, 28, 44, 46, 50, 53, 56, 73, 86, 116, 118, 119, 121, 122, 124, 125, 133, 135, 156
molecules, 14, 61, 64, 86, 88
monitoring, 39, 160

monoclonal antibody, xi, 84, 131, 147, 157, 159
monomers, 57, 61, 65
morphology, x, 12, 14, 22, 49, 100, 103, 105, 107, 110, 113, 124, 138, 142, 143, 147, 149, 158
mortality, 49
motion, 145
motor neuron disease, 39, 72
motor neurons, ix, 21, 29, 32, 37, 39, 41, 46, 50, 51, 59, 77, 78, 99, 102, 103, 137, 138, 157, 158
motor system, 29, 162
movement, ix, 5, 23, 50, 54, 59, 66, 69, 78
mRNA, 32, 36, 41, 44, 45, 91, 96, 117, 129, 135, 157, 161
mucosa, ix, x, 99, 100, 101, 102, 105
multiple sclerosis, vii, viii, ix, 26, 27, 39, 50, 81, 82, 83, 84, 85, 86, 87, 88, 89, 92, 93, 94, 95, 96, 134
muscles, 32, 55
mutagenesis, viii, 53
mutant, 15, 32, 37, 44, 58, 59, 62, 63, 70, 72, 76, 78, 88, 89, 127
mutation, ix, 35, 40, 44, 46, 47, 49, 51, 54, 67, 69, 72, 74, 75, 79
myelin, ix, 3, 8, 15, 34, 39, 48, 49, 51, 72, 81, 82, 83, 84, 86, 88, 89, 90, 91, 92, 93, 95, 96, 97, 133, 134, 156
myelin basic protein, 39, 82, 89
myosin, 31, 48

N

necrosis, 84
needs, 10, 126
neocortex, 155, 157, 158
neonates, 38
nerve, vii, ix, 5, 6, 8, 10, 13, 15, 16, 18, 20, 21, 23, 31, 33, 36, 37, 47, 57, 58, 59, 62, 65, 66, 68, 71, 73, 76, 78, 82, 91, 93, 96, 99, 100, 102, 105, 111, 112, 113, 116, 117, 125, 126, 127, 128, 130, 146, 156, 158, 159
nerve conduction velocity, 62, 117, 126, 127
nerve fibers, 23, 58, 71, 73, 78, 156, 158
nerve growth factor, 16, 20, 23, 82, 91, 93, 96
nervous system, ix, xi, 6, 12, 21, 43, 67, 69, 74, 77, 86, 87, 89, 90, 99, 100, 111, 112, 113, 114, 125, 127, 129, 131, 133, 135, 156, 161
network, viii, 3, 4, 29, 45, 53, 54, 59, 61, 65, 66, 67, 68, 73, 75
neural connection, 153
neuroblasts, 28
neurodegeneration, ix, 54, 65, 66, 68, 69, 82, 93

neurodegenerative disorders, 125
neurofibrillary tangles, vii, 10, 19, 36
neurofilaments, vii, viii, ix, x, xi, 2, 3, 5, 6, 8, 10, 12, 13, 14, 15, 17, 18, 19, 20, 21, 23, 25, 26, 27, 41, 42, 43, 44, 45, 46, 47, 48, 49, 50, 51, 54, 55, 56, 58, 59, 60, 63, 67, 69, 70, 71, 73, 74, 76, 77, 78, 79, 81, 85, 86, 87, 88, 90, 92, 93, 95, 100, 117, 124, 125, 129, 131, 133, 135, 155, 156, 157, 158, 159, 160, 161, 162
neurogenesis, 19, 42, 47, 71
neurokinin, 113
neurological disability, 86
neurological disease, viii, 10, 25, 31, 37, 39, 41, 69
neuromyelitis optica, 84
neuronal cells, 2, 6, 38
neurons, vii, viii, ix, x, xi, 1, 2, 3, 4, 5, 6, 7, 9, 10, 11, 12, 14, 15, 17, 18, 19, 20, 21, 22, 23, 25, 26, 28, 29, 30, 31, 32, 33, 35, 36, 37, 41, 43, 45, 46, 47, 48, 50, 51, 53, 54, 55, 58, 59, 60, 62, 63, 64, 66, 70, 71, 73, 74, 75, 77, 86, 91, 94, 96, 99, 100, 101, 102, 103, 105, 106, 107, 108, 109, 110, 111, 112, 113, 115, 116, 117, 119, 120, 123, 124, 125, 126, 128, 129, 131, 133, 134, 136, 137, 138, 139, 140, 141, 143, 144, 145, 146, 147, 148, 149, 150, 151, 152, 153, 155, 156, 157, 158, 160, 161, 162
neuropathic pain, 47
neuropathy, viii, 25, 36, 37, 41, 42, 45, 51, 63, 66, 67, 70, 71, 75, 79, 116, 117, 124
neuropeptides, 111
neuroprotection, 90
neurotoxicity, 41, 87
neurotransmission, 101
neurotransmitter, 3, 87
neurotrophic factors, 12, 14, 16, 17, 18, 22
nitric oxide, 82, 86, 87, 88, 92, 93, 94, 95, 96, 97
nitric oxide synthase, 93
node of Ranvier, 85
nodes, 30, 83, 85
noise, 2, 9, 14, 19, 21, 23
nonsense mutation, 75
normal aging, 66
nuclei, 137, 138, 140, 141, 146, 149, 154, 155, 157, 158, 159, 162
nucleus, 132, 133, 135, 136, 137, 140, 142, 143, 144, 145, 146, 148, 149, 150, 154, 155, 156, 157, 158, 159, 160, 161

O

obligate, viii, 28, 46, 53, 56, 68, 74

observations, ix, 23, 40, 54, 65, 69, 91, 102, 105, 112, 146
occipital lobe, 150
oculomotor, xi, 131, 132, 137, 138, 140, 156
oculomotor nerve, 137
Oklahoma, 115
oligodendrocytes, ix, 30, 81, 83, 84, 85, 86, 88, 90, 91, 92, 94, 97
oligodendroglia, 48, 96
oligomers, 5, 51, 60, 61, 78
optic nerve, 21, 22, 36, 63, 64, 75, 77, 90, 91, 94, 135, 155, 157, 161
organ, 3, 7, 8, 9, 10, 14, 23
organelles, 2, 26, 31, 40, 77
organization, viii, 19, 25, 45, 48, 54, 57, 62, 66, 67, 75, 76, 90, 103, 113, 133, 136, 138, 140, 145, 154, 155, 156, 157, 158, 159, 160
osmium, 9, 15, 16
output, ix, 99, 140, 156
oxalate, 21
oxidative stress, 31

P

pancreas, 100
parallelism, xi, 131
paralysis, 32, 65
parietal cortex, 152
Parkinson's disease, 27, 33, 34, 35, 63, 66, 67, 134
particles, 2, 26, 31, 40
partnership, viii, 53, 54, 64, 65, 68, 69
parvalbumin, 140, 156
passive, 9
pathogenesis, 32, 37, 54, 66, 82, 84, 94, 96
pathology, ix, 2, 10, 21, 37, 47, 48, 49, 66, 81, 82, 85, 86, 92, 96, 118, 135, 158
pathophysiology, 82, 90, 157
pathways, 3, 12, 32, 60, 92, 95, 97, 117, 161
peptides, 134
periodicity, 90
peripheral nervous system, 32, 33, 42, 67, 89, 125, 127
peripheral neuropathy, 117, 124
perspective, 48
PGN, 132, 149
pH, 118
phenotype, ix, 33, 81, 89, 90, 91
phenylalanine, 39
pheochromocytoma, 17
phosphates, 62

phosphorylation, vii, viii, ix, x, xi, 2, 8, 10, 15, 16, 17, 18, 19, 20, 23, 25, 26, 27, 29, 30, 31, 33, 34, 35, 37, 38, 42, 43, 44, 45, 46, 47, 48, 49, 50, 51, 53, 56, 62, 64, 68, 69, 71, 72, 73, 76, 78, 82, 86, 87, 88, 89, 90, 92, 95, 96, 115, 116, 117, 119, 120, 121, 122, 123, 124, 125, 126, 127, 128, 129, 130, 133, 134, 135, 155, 159, 160, 161, 162
photographs, 147, 152
photomicrographs, 138, 139
pigs, x, 9, 12, 14, 19, 20, 21, 100, 135
plaques, 36, 46, 82, 88, 95
plasma, 85, 89, 150
plasma membrane, 89, 150
plasticity, 10
plexus, x, 100, 101, 102, 103, 105, 106, 110, 111, 112, 113, 114, 126
PLP, 82, 89
PM, 40, 41, 48, 50, 70, 77, 79, 94
point mutation, 35, 66
polymerization, viii, 25, 59
polymers, viii, 25, 29, 60, 61, 135
polymorphism, 48
polypeptide, 5, 56, 72, 135
poor, 40, 83, 85, 88, 149
population, vii, 2, 3, 5, 18, 44, 51, 78, 103, 106, 107, 116, 153
potassium, 91
power, xi, 131, 138, 139, 147, 148, 152
precursor cells, 83, 91, 92
prediction, 39, 40
pressure, 36
prevention, 47
primary progressive multiple sclerosis, 95
primary visual cortex, 153, 156, 160
primate, xi, 131, 142, 143, 145, 153, 155, 160
principle, 140
production, x, 8, 91, 92, 115, 116, 126
progesterone, 119
program, 9
progressive supranuclear palsy, 63, 66
proliferation, 93
propagation, 83
prosthesis, 2
protein family, 48, 55
protein folding, 65, 72
protein kinase C, 47, 128
protein kinases, 29, 49, 77, 95, 96
protein synthesis, 20, 29, 44, 58, 72, 95, 128
proteins, vii, viii, 1, 2, 3, 5, 6, 8, 10, 14, 15, 16, 17, 20, 21, 22, 25, 26, 27, 28, 30, 31, 35, 36, 37, 40,

42, 44, 46, 47, 48, 49, 50, 51, 53, 54, 55, 56, 58, 59, 61, 62, 64, 65, 67, 68, 69, 70, 72, 74, 75, 77, 78, 89, 96, 112, 117, 118, 127, 130, 134, 135, 136, 154, 155, 156, 157, 158, 159, 160, 161
proteoglycans, 147
proteolipid protein, 82, 89, 97
proteolysis, 8, 10, 20, 44, 87, 95
pulse, 29, 59, 63
pyramidal cells, 151, 152, 153

Q

quality control, 65

R

rain, 47, 94
range, 6, 40, 93
reading, 154
reality, 14, 92
reasoning, 14
receptors, 3, 17, 96, 127, 128
recognition, 133
recovery, 22, 48, 92
recycling, 8
redistribution, 83
reduction, vii, 9, 10, 11, 13, 15, 17, 37, 47, 58, 88, 116, 117, 124
reflexes, 103, 105
regenerate, vii, 1, 2, 10, 14
regeneration, 2, 10, 13, 14, 17, 18, 19, 21, 22, 37, 135, 155, 157, 161
regrowth, 18
regulation, 26, 30, 31, 40, 41, 47, 58, 77, 89, 125, 126, 157
regulations, 31
regulators, 17
relapses, 83, 84
relationship, 58, 69, 86, 87
relationships, 113
relevance, 50
remission, 83
remyelination, 83, 85, 92, 94, 96
repair, 61, 85, 95, 97
replacement, 20, 48, 76
residues, 30, 31, 49
resistance, 83
resolution, 7, 83
respiration, 87, 93

reticulum, 158
retina, 71, 135, 136, 156, 158, 161, 162
retinoblastoma, 135, 158
rhythm, 150
rings, 146
risk, 32, 49, 66, 116, 117, 127
risk factors, 32
rodents, 138, 140, 143, 146, 150
room temperature, 120
Rouleau, 32, 43, 49, 72

S

sample, 103
sclerosis, ix, 81, 82, 83, 84, 86, 88, 93, 94, 95
search, 84, 86
secrete, 91, 119, 123, 124
secretion, 126
selenium, 119
self, 9, 93, 105, 133
self-destruction, 9
sensorineural hearing loss, 2, 8, 9, 10, 12, 15, 22
sensory systems, 155
series, 91, 125, 150, 153
serine, 30, 87, 116, 118, 121, 122, 125
serum, x, 40, 84, 115, 119, 120, 123, 124, 125, 126, 129
SGP, 132, 138, 139, 140
shape, viii, 2, 10, 25, 26, 31, 109, 143
shaping, 21, 46, 158
sheep, 105, 106, 107, 111
shock, 37, 40, 67, 69, 71
siblings, 36, 45
sigmoid colon, 111
sign, 85
signal transduction, 21
signaling pathway, 22, 87
signalling, 92, 97
signals, vii, 1, 15, 16, 19, 83, 91
silver, 26, 54, 102, 103, 105
similarity, 140, 158
sites, 17, 27, 29, 30, 42, 46, 51, 62, 64, 72, 73, 87, 89, 94, 117, 118, 125, 126, 129, 160, 162
skeleton, 133, 153
skills, 36
small intestine, 107, 110, 111, 112, 113, 114
smooth muscle, 103
sodium, 83, 85, 94, 95
spastic, 94
specialization, 75

species, x, 6, 64, 99, 101, 102, 106, 107, 134, 137, 138, 140, 142, 145, 146, 149, 150, 159, 161
specificity, 133
speed, 29
sphincter, 100
spinal cord, 21, 32, 47, 58, 59, 77, 78, 86, 93, 95, 137
spindle, 26
spine, 153
stability, 2, 6, 8, 13, 18, 31, 40, 67, 87, 95
stabilization, 31, 72
stages, 5, 6, 83, 88, 89, 90, 92, 95, 96
standard error, 119
stellate, 151, 152
stem cells, 46, 55
stimulus, 105
stoichiometry, 22, 29, 31
strabismus, 156
strain, 78
strength, 26
stress, 17, 44, 45, 62, 68, 117
stroke, 116
structural changes, 17
structural characteristics, 100, 103
structural protein, 2, 31, 103
subarachnoid hemorrhage, 39
submucosa, x, 100, 101, 102, 108
substrates, 50, 91
Sun, 23, 30, 50, 78, 157
supply, 61
suprachiasmatic nucleus, 132, 136, 159, 162
surface area, 91
survival, vii, 1, 16, 18, 19, 20, 21, 23, 41, 70, 88, 89, 90, 91, 92, 93, 94, 96, 97, 155
susceptibility, 8, 22, 36, 48, 67, 158
suspects, 46
suture, 145
swelling, 45, 87, 90, 147
Switzerland, 25
symptoms, 92
synapse, 3, 14, 68
synaptic plasticity, 26, 87
syndrome, 39, 44
synthesis, viii, 10, 29, 41, 53, 61, 62, 116, 129
systems, 29, 112

technical assistance, 110, 154
tensile strength, viii, 2, 26, 40, 53, 55, 134
terminals, 3, 62, 105, 146, 159, 160
thalamus, 141, 144, 146, 147, 148, 154, 157
theory, 6, 54, 92
therapeutic agents, 83
therapy, 22, 82, 85, 95
thinking, 36
thresholds, 14, 17
time, x, 5, 9, 12, 29, 31, 32, 40, 83, 90, 92, 95, 100, 120, 125, 153
time periods, 9
timing, 136, 153
tissue, 7, 8, 47, 84, 85, 86, 88, 100, 101, 103, 110, 124, 130, 134, 150
TNF, 84
toxicity, 19, 37, 47, 50, 78, 95, 120, 123
toxin, 79, 84, 128
traffic, 70
trajectory, 14
transcription, 17, 133
transduction, vii, 1, 3, 41
transection, ix, 81, 82, 87, 91, 94, 96, 156
transferrin, 119
transgene, 44, 62, 72
transition, 5, 14
translation, 65
translocation, 29, 49, 59, 94, 160
transmission, 2
transport, viii, ix, xi, 2, 10, 19, 20, 21, 22, 25, 26, 29, 30, 31, 32, 34, 37, 40, 41, 43, 45, 47, 48, 49, 50, 51, 53, 54, 55, 56, 58, 59, 60, 61, 62, 63, 64, 66, 67, 68, 69, 70, 71, 72, 73, 74, 75, 76, 77, 78, 79, 81, 85, 87, 89, 94, 116, 117, 125, 127, 128, 135, 154, 158, 159, 160, 161
transportation, vii, 1, 31
trauma, 2, 9, 11
tremor, 89
triggers, 8, 66
trypsin, 119
tumors, 101
turnover, 49
type 1 diabetes, xi, 116, 129
tyrosine, x, 49, 115, 118, 120, 121, 122, 124, 125, 128

T

targets, 70
tau, 36, 44, 49, 50, 67, 77, 116, 117, 128, 129

U

UK, 81, 82
underlying mechanisms, 38

uniform, 12, 104
urea, 41
urine, 118, 122

V

variability, 37, 84, 140, 151
variable, 103
velocity, viii, ix, xi, 17, 25, 30, 37, 54, 58, 65, 69, 73, 76, 78, 79, 87, 116, 117, 125, 126, 134
vertebrates, 58, 135, 161
vesicle, 5
vision, vii
visual area, 152
visual system, xi, 131, 135, 141, 153, 155, 157
visualization, 104, 105, 107, 133

W

white matter, 38, 85, 86, 95
wild type, xi, 116, 118, 121, 122, 124
wind, 56
withdrawal, 117
women, 36
words, 88
work, 36, 37, 39, 40, 69, 83, 85, 86, 87, 110, 134
workers, 91

X

X chromosome, 89

Y

yield, 91
yuan, 53

Z

zinc, 44